Bogs & Fens

Bogs
&Fens

Ronald B. Davis

A Guide
to the
Peatland Plants
of the
Northeastern
United States
and Adjacent Canada

University Press of New England | Hanover and London

University Press of New England
Manufactured in the United States of America
Typeset in Scala

For permission to reproduce any of the material
in this book, contact Brandeis University Press,
415 South Street, Waltham, MA 02453,
or visit brandeisuniversitypress.com

Library of Congress Cataloging-in-Publication Data

Davis, Ronald B., author.
Bogs and fens: a guide to the peatland plants of the
northeastern United States and adjacent Canada /
Ronald B. Davis.
        pages   cm
Includes bibliographical references and index.
ISBN 978-1-61168-793-4 (pbk.) — ISBN 978-1-61168-
948-8 (ebook)
1. Peatland plants—Northeastern States—Identification.
2. Peatland plants—Canada—Identification.
3. Bogs—Northeastern States. 4. Bogs—Canada.
5. Fens—Northeastern States. 6. Fens—Canada. I.
Title.
QK938.P42D38 2016
577.68'7—dc23   2015034458

5   4   3

Contents

Preface

The idea for this book developed over the past decade, while I was volunteering as a docent and guide at the Orono Bog Boardwalk (cover photo) in Bangor and Orono, Maine. It became apparent to me that the many people who visit peatlands (bogs and fens) in this region need a book on peatland plants to enhance their outdoor experience, and I vowed to write one. During these same years I took up the hobby of nature photography and started to collect the plant photos for the book. Prior to these "retirement" activities, I had been a professor of biology and Quaternary studies at the University of Maine, and before that a biology professor at Colby College. Over those forty-three academic years I led many class field trips to bogs and fens. Additionally, in the period 1980 to 2003 I directed a program of ecological research on the peatlands of Maine, and also visited many peatlands elsewhere in northeastern United States and other parts of the world. A few of the publications that resulted from that work are listed later in the book.

My enthusiasm for bogs and fens was kindled much earlier by Paul G. Favour Jr., park naturalist in the 1950s and '60s at Acadia National Park. During my summers at Acadia as a ranger naturalist, Paul was my boss and friend. One day, he guided me to a coastal bog at Acadia, the Big Heath, and introduced me to the wonders of peat bogs, a seminal occasion for me. Much later at the University of Maine, my colleague Dennis S. Anderson and I worked together on research on the ecology of the bogs and fens of Maine. This book is partly an outgrowth of that productive collaboration. A draft of this book was reviewed by Dennis, and also by Albert Larson and Glen H. Mittelhauser, all of whom made useful suggestions.

I took a great majority of the photos in this book. The others were contributed by 22 people. Without their photos, this book would not have been possible, and I gratefully acknowledge their contributions. Most of these photos were contributed by the following people: Glen H. Mittelhauser, Marilee Lovit, Russ Schipper,

Robert W. Smith, Donald S. Cameron, Arthur Haines, and Anton A. Reznicek. Smaller numbers of photos, but no less important ones, were contributed by Allen Chartier, Joshua Fecteau, George W. Hartwell, Louis M. Landry, Steve Matson, Keir Morse, Bruce Patterson, Graham R. Powell, Robert Routledge, Hermann Schachner, Marilylle Soveran, Ellis Squires, Greg Vaclavek, Warren H. Wagner Jr., and Leanne Wallis. All these contributors are credited in the captions of their photos. Anton A. Reznicek was additionally helpful in expediting contacts with some of these photographers. I am grateful to David d'Entremont for pointing out a misidentification of a blackberry species in the first printing of this book, and have corrected the error for the second printing. Other people too numerous to mention supplied essential information for the part of this book that lists peatlands with boardwalks in the 19 states and provinces covered, for which I am grateful.

Finally, my wife, Lee, has supported my efforts over the years as I gathered photos for this book. She provided occasional field assistance, and in recent years has tolerated the long hours I have spent at my computer archiving and editing photos and preparing this text. Additionally, she did a careful edit of the final manuscript. I thank her and all the above-mentioned people for their input and encouraging support.

Introduction

This book is about plants and places. The places are wet, soggy, and difficult to visit. Until recent decades, with the construction of boardwalks to provide ease of access, only small numbers of intrepid rubber-booted hunters, naturalists, and biology professors along with their not-always-so-enthusiastic students had experienced the challenging terrain. These places are bogs and fens, the two major categories of peatlands. Peatlands are a major category of wetlands. They are underlain by water-saturated peat, in some places deep enough to swallow a tour bus. When I started studying such places in the 1960s, I was told by my friends to wear my broad-brimmed ranger hat, so if I didn't return they could go out to find where I sank! Actually, there is little such danger, and since my initial outings there has been a huge increase in the number of people visiting peatlands, not by walking on, and damaging the plants and the squishy peat, but by way of boardwalks.

Over the past several decades, a burgeoning interest in outdoor activities, a desire to get out in nature to see wildflowers, birds, and other wildlife, and a growing concern about the environment have led to the establishment of large numbers of parks and reserves where people may enjoy and learn about the natural world. The fact that many of these parks and reserves protect valuable wetlands has led to the construction of boardwalks to enable individuals, families, and school groups to experience the beauty and learn about the natural history of these soggy places without getting their feet wet. Boardwalks have been established in over 100 wetlands, most of them peatlands, in the area covered by this book (map 1). Judging from the visitation figures given to me by agencies managing some of these facilities, and a conservative extrapolation of these figures to the entire region, I can safely say that several hundred thousand visits are made each year to bog and fen boardwalks in northeastern United States and adjacent areas of Canada. Unfortunately, there has been no guidebook for use at the region's peatlands that focuses on the plants and their habitats. It has been my goal to write one, and here it is!

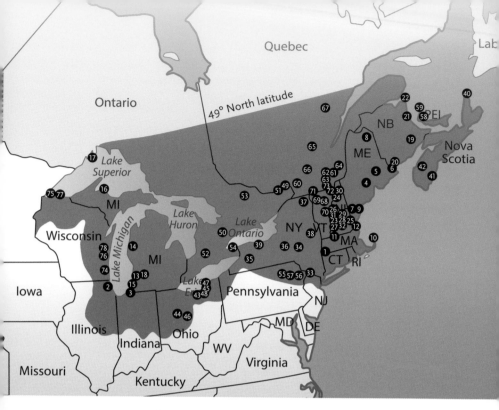

Map 1 Map of area (green) covered by this book. The edge indicating the southern limit from Massachusetts to Illinois, western limit in Illinois and southern two-thirds of Wisconsin, and southern limit in northern Wisconsin represents the terminal position of the most recent (Wisconsinan) continental ice sheet. The northern limit in Ontario and Québec Provinces is 49° N. lat., with the exception of a more northern sliver of the Gaspé Peninsula. The remaining limits follow the northwestern shore of Lake Superior and state and provincial boundaries and shorelines. The numbers indicate the positions of the 78 peatlands with boardwalks that are described in the final part of the book.

When out on a peatland boardwalk, you may easily use this book to enhance your experience of the environment and plant life around you. Along with this book, don't forget to bring your hand lens and/or binoculars, camera, insect repellent, drinking water, and a snack (nutty chocolate is highly recommended by this author)! You will also find this book helpful after you return home to look at your photos. "What is that plant in the photo with the spectacular red flower on a long stalk?" Let's look it up in the book.

Although the book enables you to identify the plants, it is more than that. It is an eco-botanical guide. By "eco-botanical" I mean that it aims to increase your understanding of the ecology of the plants and the peatlands where they grow. These plants are well adapted for growth in peatlands. What are some of these adaptations, and exactly what are they are adapting to?

No botanical or ecological expertise is needed to use the book. Following this introduction, I present some fundamental botanical concepts and define some of the terms that are used in the species descriptions. I also define ecological and landscape terms such as "wetland," "peat," "bog," and "fen," among others, that are important for understanding the plants' environments. Throughout this part of the book, terms that may be new to you, and that are important for understanding what comes later, are given in italics. More information on peatland plants and the peatlands where they live may be found in the books and articles I list in the Further Reading section. Authors' names and publication dates cited anywhere in the text refer to the publications in that list.

Just prior to the Species Description section that makes up the heart of this book, I include a Plant Structure Terminology section that defines terms used to describe plant structures such as types of leaves, flower parts, and fruits. These definitions will help you to distinguish one plant species from another. Each species description includes one or more photographs, a brief verbal description to aid in identification, and information on where the species may be found.

The Species Descriptions section features 98 of the most common and widespread plant species of the peatlands of our region.

These descriptions include information on additional species that are less common or widespread in our peatlands but may resemble the featured ones, so you can tell them apart. Thirty-four such "comparative species" are distinguished. After the descriptions, I include an Additional Peatland Species section, where I list and briefly describe 23 other species you might encounter, and where you can obtain more information about them.

In the concluding part of the book, Visiting Peatlands with Boardwalks, I direct you to peatlands in each state and province (map 1) where you may easily put this book to good use. I list 78 such peatlands, briefly characterize each one, and give current Internet sources where you may find additional information including travel directions.

The Plants of Peatlands

This book deals only with *vascular plants*, namely plants with internal, specialized liquid-conducting tissues. Vascular plants include all the seed-producing plants with flowers, fruits, or cones, and some plants lacking these reproductive structures such as ferns and horsetails. Not included are the mosses and liverworts, small nonvascular plants that are abundant in most of our peatlands. Peat mosses (*Sphagnum* spp.) form near-continuous moss carpets at parts of many of these wetlands. There are many species of peat mosses, and they have a major influence on the environment of the vascular plants. The identification of these important plants is a subject for another book. Books that have already been written about them include McQueen (1990) and Crum (1992), among others.

I give both the American common name and the Latin binomial name of each of the described vascular plant species. The first part of a binomial name is the *genus* (pl. *genera*) to which a species belongs, for example, *Larix*. The second part is the species within that genus, for example, *laricina*. For a few species, I use a Latin trinomial name. In such cases, the third term designates a variety (var.) or a subspecies (ssp.) of that species. When I use "spp." After a generic name, as in *Carex* spp., it signifies multiple unspecified species of the genus. Latin generic, binomial, and trinomial names are given in italics.

There is a good deal of regional and local variation in the common names of plant species; for example, the tree species *Larix laricina* may be called hackmatack, tamarack, or American larch. The Latin name is much more of a constant throughout the range of a species, but it may change over time as botanists learn more about the genus and species. For these reasons, for the Latin names of species I follow a single botanical reference, Haines (2011). For common names I also follow Haines, although occasionally I depart from his hyphenations in favor of generally accepted usages that will be more familiar to the reader.

The *family* of each of the 98 featured species is indicated in its description. Some families of vascular plants are particularly well represented in our bogs and fens, most notably the *heath family* (Ericaceae) and the *sedge family* (Cyperaceae). Thirteen of the ericaceous species that are widespread in our peatlands are featured in this book. Additionally, in some of these descriptions, I give distinguishing characteristics of other, less-common ericaceous species of our peatlands that might be confused with the featured ones.

Over 100 sedge species occur in the peatlands of our area, at least 69 of which occur in Maine peatlands alone (Anderson et al. 1996). Even experienced botanists have difficulty distinguishing some sedge species from one another, and these species are particularly challenging for the casual naturalist. For these reasons, I feature in this book only 14 of the most common, widespread, or distinctive sedge species and give information for distinguishing them from other species that might be confused with them. Several more sedge species that are widespread but not as common in our peatlands are listed in the Additional Peatland Species section. All the sedge species that are common in bogs and fens across our area are illustrated by photographs in a recently published guide to the sedges of Maine (Arsenault et al. 2013).

Sedges superficially look like grasses, but the two families are not closely related. You will find information for distinguishing grasses from sedges on the Canada reed grass page, and information on the major characteristics of sedges on the boreal bog sedge page.

The three-volume comprehensive illustrated flora of the northeastern United States and adjacent Canada by Gleason (1968) is especially helpful as a reference on plant identification, but some of the Latin names have changed since 1968. More current names are given by Haines (2011).

Plant Occurrence and Peatland Habitats in the States and Provinces

The occurrence section of each species description includes the species' geographic distribution within, and in some cases beyond, the area covered by the book. I follow a single source for species presence or absence in each of the 19 states and provinces, namely, the U.S. Department of Agriculture Plant Profiles or Fact Sheets, found at http://plants.usda.gov/java/factSheet.

Additionally, the occurrence section indicates the habitats where the species grows, such as alkaline fen or bog margin. These and other peatland habitats are described prior to the species descriptions. Habitat information for a species alerts the reader on where to look for it, and allows for a questioning of a species' identity when it appears to be out of place. I emphasize "questioning," not completely eliminating, the possibility of a correct identity, as plant species occasionally occur in places that are atypical for them.

Wetland Indicator Status of Plant Species

When you encounter a plant in a squishy peatland, and it looks very much like a familiar plant you have seen many times in the forests or fields around your house, you may wonder whether the plants are really the same species. Are you missing some key feature that distinguishes them as different species? Or does the same species grow in both places, despite the stark environmental differences? In the identification of peatland plants, you will sometimes find it helpful to know the extent to which a species is confined to wetlands. For this reason, and others, as will become apparent, in each species' description I give you its *wetland indicator status* (wis).

I obtained these wis designations from an extensive list of plant species that has been developed by U.S. government agencies for all regions of the United States. I use the wis designations for the combined "Northcentral and Northeast Region" from the 2012 version of the list (http://plants.usda.gov/wetinfo.html). The original purpose of the list was to provide a tool for recognizing and delineating wetlands, so that laws designed to protect wetlands could be properly applied. However, in this book I use wis designations to describe the extent to which each plant species is confined to wetlands. Here is how it works.

A plant species may be either an *obligate* wetland species, meaning that it almost always grows in wetlands (estimated at more than 99 percent of the time), or a *facultative* wetland species (grows in both wetlands and non-wetlands). For example, bog rosemary (*Andromeda polifolia* var. *glaucophylla*) is an obligate (OBL) wetland species. As the extent to which facultative wetland species occur outside wetlands differs greatly from species to species, the facultative condition has been split into three categories: (1) *facultative wetland* (FACW), where the species usually occurs in wetlands (est. 67–99 percent of the time) but also in non-wetlands; (2) *facultative* (FAC), the species occurs in wetlands (est. 34–66 percent of the time) but also in non-wetlands; and (3) *facultative upland* (FACU), the species usually occurs in non-wetlands (est. 67–99 percent of the time) but also in wetlands. Examples of FACW, FAC, and FACU species are, respectively, black ash (*Fraxinus nigra*), black chokeberry (*Aronia melanocarpa*), and black huckleberry (*Gaylussacia baccata*). Obligate *upland* species (UPL) almost never (est. less than 1 percent of the time) occur in wetlands.

Plant Growth Forms

The plants in this book are grouped by *growth form* or habit, sometimes called life form. The outer edges of the pages are color coded according to plant growth form, so you may more easily find the page with the plant you are looking for. The first question to ask yourself when identifying a plant is, what is its growth form? Definitions of the various growth forms vary in the literature, and for this reason I give in the following paragraphs the definitions used in this book.

In our peatlands, there are several growth forms of *woody plants*. These plants have live woody stems aboveground all year long. In addition, needle- and broad-leaved evergreen woody plants retain live leaves in winter. Whether evergreen or deciduous, in the growing season these plants develop new buds on their woody stems that will overwinter and open in the next spring to form new branches, leaves (called needles in some conifers), and flowers or cones. Apart from the extensive carpet of peat mosses on the surface of some of our peatlands, and apart from sedge-dominated fens, woody plants provide most of the plant cover and dominate the vegetation of our peatlands. In this book, the following definitions are used for the growth forms of woody plants.

Trees are woody plants that typically have a single main stem, the trunk, and when fully grown under favorable conditions in our region are more than 20 ft. (about 6 m) tall. However, under unfavorable conditions, as near tree line high on a mountain, or at the central area of a bog, tree species may grow so slowly that they never reach 20 ft. in their lifetimes. Another exception is that some tree species can sprout from their bases when the aboveground part of the tree has been killed by a natural event or cut by a logger, and such growth may produce a tree with multiple trunks, as is often the case for red maple.

Tall shrubs are woody plants that have multiple stems arising from the ground, and when fully grown under favorable conditions are typically taller than 40 in. (about 1 m) and shorter than 20 ft. (about 6 m). Under unfavorable conditions, tall shrubs (e.g., mountain holly) may not reach 40 in. in height when fully grown.

Short and *dwarf shrubs* are woody plants that typically have multiple stems arising from the ground, and when fully grown under favorable conditions are less than 40 in. (1 m) tall, frequently less than 20 in. (0.5 m). As with trees and tall shrubs, the maximum height reached varies with growing conditions. Most shrub species in the central portion of bogs in our region have this growth form.

Prostrate (trailing) shrubs are woody plants that trail along the ground. Some stems may grow upward at their tips, but rarely by more than 4 in. (about 10 cm). Now let's switch from woody to herbaceous plants.

Herbaceous plants or *herbs* are nonwoody flowering plants that die down to the ground at the end of the growing season in our region. The *graminoid* plants with grasslike leaves, namely the grasses, sedges, and rushes, are sometimes put in a growth form all their own, but here they are combined with other herbaceous plants. All but a small number of herbaceous species in our undisturbed peatlands have live overwintering parts at or below ground surface, which sprout new aboveground parts in the spring. These species are *perennials*. A few *annual* species, those that overwinter only as seeds, may be present at peatland edges or disturbed peatland surfaces. Peatland-edge annuals include halberd-leaved tearthumb and devil's beggar-ticks.

Ferns and *horsetails* are vascular plants lacking flowers, fruits, and seeds. Ferns have leaves called *fronds*. In most species of our area, fronds are subdivided to take on a lacy look. Ferns reproduce by release of spores from small patches on the fronds called *sori*. The spores germinate on moist surfaces and develop into tiny plants called gametophytes that reproduce sexually by motile sperm that fertilize eggs. It is that fertilized egg that develops into the first-year fern. All ferns in the peatlands of our area are perennials that die down to the ground at the end of the growing season and overwinter as *rhizomes* (modified stems underground or partly buried at ground surface). In the spring the rhizomes sprout new fronds that unfurl as fiddleheads. All fern species have fiddleheads. In our area, the fiddleheads of ostrich fern or "fiddlehead fern" (*Matteuccia struthiopteris*) are eaten, and many people apply the name fiddlehead exclusively to the unfurling leaves of this species. While this species does not occur in our peatlands (it occurs in mineral-soil wetlands), many other fern species, all of them producing fiddleheads in the spring, are common at many of our peatlands. None of these fiddleheads are considered edible, and some may be harmful to eat. Information on horsetails is given later on the woodland horsetail page.

The species descriptions are arranged by growth form in the same order as in the prior several paragraphs. Within each growth form, the species are ordered alphabetically by common name.

The Ecology of Peatlands

In this section I present concepts and definitions that pertain to bog and fen environments and to adaptations of plants for survival in these environments.

Wetlands Defined

As peatlands compose a major category of wetlands, an understanding of the factors that differentiate wetlands from other ecosystems will help the reader to understand what peatlands really are, and to appreciate the environmental challenges that must be met to be a successful peatland plant. Although the term *wetland* has been defined in several ways (Keddy 2010; Mitsch and Gosselink 2015), for the purposes of this book I use three criteria to define it: (1) wetlands are ecosystems with a water table at or very near the soil surface (water-saturated soils) for most or all of the growing season, and commonly throughout the year; (2) wetland soils below some minimal depth lack dissolved oxygen because of two factors: (a) soil spaces or pores that typically would be open to air circulation in non-wetland soils are filled with water, and (b) decay of organic matter depletes dissolved oxygen faster than it can be replenished because of 2a; and (3) a majority of the dominant plant species in the ecosystem are well adapted for establishment and growth under the conditions described in criteria 1 and 2. Such species are called *hydrophytes*. In wetland delineation procedure in the United States, more than 50 percent of a site's dominant species must be in OBL (obligate), FACW (facultative wetland), or FAC (facultative) categories (Mitsch and Gosselink 2015).

Wetland Indicator Status and Growth Form

Many readers of this book will have observed that when an area of forest in our region is artificially flooded, as by road construction, most or all of the trees are killed, and in the ensuing years are replaced largely by herbaceous plants including cattails and sedges, in some instances including shrubs, to form a marshy wetland. This suggests there are herb and shrub species in our region that are relatively tolerant to wetland conditions compared to most of our trees.

A tally of the WIS categories of the 155 peatland species named in this book indicates that only one, or 5 percent, of the 20 tree species is OBL, and the remainder near-equally distributed among the three facultative categories. The 45 shrub species are more heavily weighted toward OBL, with 31 percent OBL, 40 percent FACW, 16 percent FAC, 9 percent FACU (facultative upland), and 4 percent UPL (upland). The 90 herbaceous species are even more heavily weighted toward OBL, with 81 percent OBL, 14 percent FACW, 3 percent FAC, and 2 percent FACU. Of these herbs, all 32 sedge species are OBL.

These numbers indicate that the woody species of our peatlands are much less restricted to wetlands than the herbaceous species, especially so for the trees. Nevertheless, woody species are abundant in our peatlands and dominate many of the plant communities. What survival strategies allow these species to widely occur at both uplands and wetlands? And what survival strategies allow so many herbaceous species to concentrate their populations at areas with water-saturated and anoxic soils?

Adaptations of Plants to Environmental Stresses in Peatlands

Here I describe some adaptations of vascular plants for survival in the water-saturated, anoxic, and infertile soils of peatlands. There are many ways that plants deal with these stresses, and they generally fall into two categories of adaptation: (1) anatomical-physiological and (2) avoidance. For more complete coverage of this subject see Keddy (2010) and Mitsch and Gosselink (2015).

ANOXIC SOILS AND PEATLAND MICROTOPOGRAPHY

The lack of oxygen in wetland soils presents a major challenge for survival of vascular plants. The roots of these plants must have oxygen to efficiently carry out their functions of uptake of water and mineral nutrients. Many of the herbaceous species of our peatlands possess internal air-filled vascular tissue that transports gases including oxygen from the leaves to the roots. This tissue is called *aerenchyma*. Not only does oxygen directly support root metabolism, but it leaks out of the roots into the anoxic soil to produce a narrow *oxidized rhizosphere* around the roots that detoxifies chemicals that form in anoxic conditions. Peatland species in this book that have aerenchyma include all the sedge species, podgrass (*Scheuchzeria palustris*), saltmarsh arrowgrass (*Triglochin maritima*), American bur-reed *Sparganium americanum*, and buckbean (*Menyanthes trifolia*). Additional description of aerenchyma and its functions may be found at http://soils.ifas.ufl.edu/wetlands/ teaching/Biogeo-PDF-files/Lecture-6-oxygen%20adaptations%20 2%20%5BCompatibility%20Mode%5D.pdf.

Just a short vertical distance, for example only 2 in. (5 cm), can make a big difference in the soil of peatlands and other wetlands in the concentration of oxygen, the availability of mineral nutrients, and the suitability of a site for establishment and growth of particular plant species. The presence of *hummock and hollow microtopography* with vertical relief typically ranging 2 to 20 in. (0.05–0.5 m), as in many of our peatlands, provides a range of microhabitats for plants with differing tolerances to anoxic soils. The depth of aerated soil is greater on hummocks than in hollows. The woody species of our fens and bogs have much greater success in establishing themselves and surviving on hummocks than in hollows. In doing so, they are avoiding the wettest soil with anoxia closest to the surface.

Trees in our peatlands generally have shallow root systems that are above soil water level and anoxia for much of the growing season, making the trees susceptible to windthrow. This susceptibility results in the perpetuation of hummock and hollow microtopography in wooded areas of peatlands. The root masses and trunks of the uprooted and downed trees create hummocks that are relatively dry and serve as favorable sites for future establishment of trees and other woody plants.

Sedge-meadow-type vegetation in some fens is characterized by hummocks formed by the roots and lower stems of tussock-forming sedges like tussock sedge (*Carex stricta*). These hummocks provide microsites for colonization by many species that require the drier and oxygenated soils of the hummocks (see http://www .lakeandwetlandecosystems.com/plants/grass-like-plants/tussock -sedge-carex-stricta/).

Hummocks are also present at relatively open areas of bogs and acidic fens. There they are formed by mound-building peat mosses. These hummocks provide more favorable microhabitats for woody plant establishment and survival than adjacent, wetter hollows. During your next visit to a bog or fen, notice that the woody plants thrive on the hummocks, and much less so in the wetter hollows.

LOW-NUTRIENT SOILS

The soils of our peatlands vary in the availability of mineral nutrients upon which plant growth depends. The vegetated surfaces of bogs receiving their mineral nutrients only from atmospheric precipitation and dry fallout—an "ombrotrophic" (literally, " rain fed") condition—have especially slow plant growth due to nutrient limitation. Nutrient uptake by plant roots in this environment must be especially efficient to sustain plant growth. An interesting example is provided by shrubs in the heath family (Ericaceae) like bog laurel (*Kalmia polifolia*) and leatherleaf (*Chamaedaphne calyculata*). Ericaceous shrubs are especially abundant at bogs. Fungi associated with their roots form *ericoid mycorrhizae*. It has been demonstrated that this symbiosis enhances mineral nutrient uptake and increases plant growth in nutrient-poor conditions. Additional information on these mycorrhizae can be obtained at http://en.wikipedia.org/wiki/Ericoid_mycorrhiza.

Deep rooting in anoxic soil can increase access to nutrients, especially where the peat is shallow and the deepest peat is mixed with the underlying mineral soil, as at many fens. The roots of some peatland species with aerenchyma, including tussock cottonsedge, *Eriophorum vaginatum,* are able to reach this relatively fertile layer in many of our fens. This herbaceous species can grow its roots to at least 5 ft. (about 1.5 m) depth, according to http://www.fs.fed.us/database/feis/plants/graminoid/erivag/all.html

A well-known adaptation of plants to low-nutrient status of peat soils is carnivory. This book features several carnivorous plants, including bladderworts (*Utricularia* spp.), sundews (*Drosera* spp.), and purple pitcher plant (*Sarracenia purpurea*). These plants supplement the mineral nutrient uptake of their roots by deriving mineral nutrients from their prey. More information on our carnivorous plants is given by Schnell (2002).

Peatland Vegetation

The *vegetation* at a particular place consists of all the plants growing there. Each type of vegetation has a characteristic structure and grouping of plant growth forms. There are many types of peatland vegetation, including forests, tall shrub thickets, dwarf shrub heaths, sedge meadows, peat moss lawns, and combinations of these, not to mention the aquatic vegetation of peatland pools. More specific categories of vegetation, defined largely by their most abundant species, are called plant communities (Anderson and Davis 1997, 1998). Very small peatlands may contain only one vegetation type, but most peatlands have multiple types, and each type may have more than one kind of plant community. These vegetation types with their plant communities typically are arranged in patterns that mirror environmental patterns (Davis and Anderson 2001). For example, in Maine and adjacent Canadian provinces, raised bogs commonly have concentric rings of differing vegetation (map 2). In such cases, the pattern reflects gradients of hydrology and mineral nutrients or fertility that extend from the edge to the center of the peatland, as described as follows for Orono Bog in Maine.

On the map of Orono Bog (map 2), from the periphery to the center of the bog the vegetation types are (1) *mixed wooded fen vegetation* (minerotrophic; see following Bogs versus Fens section): mixed forest of broad- and needle-leaved trees dominated by

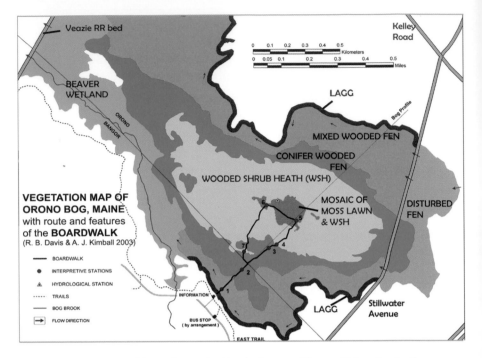

Map 2 Map of Orono Bog, a raised bog in Orono and Bangor, Maine, showing the position of the Orono Bog Boardwalk. Vegetation zones are labeled. The relatively recent beaver flowage (blue) has interrupted the concentric vegetational zones. The ombrotrophic (see text) part of the bog is outlined by the outer limit of wooded shrub heath. The outer brown band representing the lagg is diagrammatic. The lagg is actually variable in width but could not be distinguished from the next inner zone on the air photos. By Ronald B. Davis and Alan J. Kimball.

red maple trees; (2) *conifer wooded fen vegetation* (minerotrophic): predominantly coniferous forest dominated by black spruce and American larch trees; (3) *wooded shrub heath* (wsн) *bog vegetation* (ombrotrophic): nearly continuous carpet of peat moss with an abundance of dwarf shrubs in the heath family, and patches of dwarfed black spruce trees on hummocks; and (4) *mosaic of moss lawn and wsн bog vegetation* (ombrotrophic), in hollows and hummocks, respectively. Moss lawns are very wet, completely carpeted by peat moss, and have scattered sedges and dwarf shrubs of the heath family. A beaver flowage (blue on the map) has interrupted the concentric vegetation zones. Orono Bog is a gently convex bog (Davis and Anderson 2001), with a central area that is raised 2–3 ft. (0.6–0.9 m) above the bog's edge. From vegetation type 1 to 3–4, acidity increases (pH decreases) and mineral nutrient concentrations decrease in the upper peat (plant-rooting zone).

A physical feature represented on the map by a diagrammatic brown band at the bog's outer edge is the narrow, shallow moatlike area called a *lagg*. The lagg, actually variable in width, is the lowest area of the bog, and in wet seasons it carries water around it. It receives that water from two directions: from the adjacent upland, and from the raised center of the bog. Its vegetation is similar to the next inner zone of the bog, but it has greater biodiversity. Other peatlands have different vegetational patterns, as shown by Davis and Anderson (1999) on vegetation maps of 102 Maine peatlands.

Peat, and Fens versus Bogs

Peatlands, called mires or moors (with spelling variations) in most European languages, and *tourbières* in French, are distinguished from mineral soil wetlands by the presence of peat. However, many of our wetlands have only small accumulations—for example, 4 in. (10 cm) of peat atop mineral soil. Do these wetlands qualify as peatlands? What depth of peat must be present to classify a wetland as a peatland? Whatever depth is chosen will be somewhat arbitrary, as there is a continuum of peat depths in wetlands. Minimum peat depths of 8–20 in. (0.2–0.5 m), 16 in. (0.4 m) in Canada, have been used for distinguishing a peatland from a mineral-soil wetland.

PEAT

The *peat* in our peatlands is mostly water, by weight from about 95 percent at the top to about 75 percent at the bottom of the peat just above the transition to inorganic mineral matter. It seems a wonder that such watery material can support a person or a boardwalk. The support is largely provided by the rigidity of interlacing stems and branches of woody plant remains. Wet, pure peat moss peat is much less supportive. Where it is present to a substantial depth, an unwary pedestrian may sink to the waist, and a boardwalk must be supported by floats or by pilings down to mineral base.

The dry matter component of peat largely consists (by volume) of the partially decomposed remains of plants, plus a very small quantity of animal remains. Some small volume of inorganic mineral matter is always present. How much may be present for the deposit to still be considered peat? This standard by oven-dry weight has been established by the International Peat Society (IPS): the deposit may have no more than 70 percent inorganic mineral matter, or no less than 30 percent dead organic matter. The equivalent percentages by volume are very different, although they will vary, depending on the dry densities of the organic and inorganic components. I've done the calculation based on spruce wood and granite. Given that oven-dry spruce density is about 0.4 grams per cubic centimeter, and oven-dry granite approximately 2.7 grams per cubic centimeter, the dry volume of organic matter at the IPS minimum to qualify as peat would be about 75 percent. In our peatlands, except for bottom peat where it starts to transition to mineral deposit, organic matter oven-dry volumes are rarely as low as 75 percent. In recently formed peats at ombrotrophic sites, where inorganic mineral input is only from the atmosphere, dry peat volume is almost entirely organic.

The upper peat, from the surface to roughly 5 ft. (1.5 m) depth, serves as the soil or "rooting zone" for the living plants of the peatland, although few peatland species have roots deeper that 3 ft. (about 0.9 m). Peat can be many meters deeper than where the living plants are rooted, in our region to a total depth of about 33 ft. (10 m) in some peatlands. The deepest peat in most of our region's peatlands started accumulating roughly 10,000 years ago, as dated by the radiocarbon method. However, some of our peatlands have more recent origins.

Not only does peat support today's living communities; it provides an archive of life and environments of the past. Peat is such a good medium for preservation that it contains large concentrations of recognizable vegetative (leaves and stems) and reproductive (hard parts of fruits, seeds, and the walls of pollen grains) parts of plants that can usually be identified to genus and often to species. These remains are properly called fossils, despite the fact that they are not lithified. Such fossils have been used in numerous paleoecological studies. For such studies, paleoecologists use coring devices to obtain cores of peat from the surface down to the underlying mineral material. A series of levels in the core are dated by the radiocarbon method and analyzed for their fossil content. Based on multiple cores, scientists have reconstructed the stages of development of peatlands, and also the ecological histories of wider landscapes spanning many millennia (Kuhry and Turunen 2006; Martini et al. 2007; Rydin and Jeglum 2013).

FENS VERSUS BOGS

Peatlands come in a variety of types, both structural and chemical, and these types may have widely different vegetation and plant communities. A little familiarity with these types goes a long way in understanding the environments of the plants, and in forming expectations on plant species that you might encounter.

Peatlands are of two major types: *fens* and *bogs*. Fens are more abundant than bogs throughout our region. True bogs are most abundant and larger on average at the northern parts of our region. Be aware that many peatlands that are popularly called "bogs," and many that appear on maps with "bog" in their name may actually be fens. The definitions of bog and fen have varied, as I shall explain later. The definitions I use in this book are based largely on the source of mineral nutrients that the plants receive.

Although a *fen* is a wetland whose vegetation sits atop peat, some of the water that bathes the plant roots has recently been in contact with mineral soil, rock, or mineral deposit since falling as rain or snow. Fens are said to be *minerotrophic*, meaning that their plants are supplied with mineral nutrients from mineral soil, although some portion also comes from the atmosphere. Apart from rain, snow, and dustfall, in fens mineral nutrients are carried to the plants by surface and near-surface flows from an adjacent upland, from a stream that flows through the fen, from a bordering lake, by upwelling within the peat, or by some combination of these routes. The surface of a fen may be flat or gently concave.

Fens may be wooded or unwooded, or some of both. Wooded fens are sometimes called swamps, but if they are underlain by enough peat, they are more correctly called wooded fens. In this book, the term "swamp" is limited to wooded wetlands on mineral soil (therefore, not fens). Fens dominated by graminoid plants (grasses, sedges, and rushes) are sometimes called marshes, but if they are underlain by enough peat, they are more correctly called graminoid fens. These fens are most commonly dominated by sedges. In this book, the term "marsh" is limited to wetlands dominated by graminoid plants on mineral soil (therefore, not fens).

Depending on the mineral composition and structure of the surrounding and underlying mineral soils and rocks and/or the chemistry of water bodies from which a fen's waters originate, the fen's surface water can range from acidic or circumneutral to weakly alkaline. By "circumneutral" I mean a nearly neutral pH of 7, variously specified, but here meaning pH 6 to 7.5. Commonly, fens are classified into three categories depending on the chemistry of the water at the plant roots: (1) acidic fens or "poor fens" with pH 4–4.5, (2) moderately acidic fens or "intermediate fens" with pH 4.5–6, and (3) circumneutral to alkaline fens or "rich fens" with pH 6–8.5 (Anderson and Davis 1997).

Although the above fen categories are based on differences in environmental chemistry, the flora reflects those differences. The experienced peatland ecologist can recognize the three fen types from the plants that are present or absent. Even a beginner can distinguish a poor from a rich fen by their different sets of plant species. The more acidic the fen, the more its chemistry and flora resemble those of a bog (described below), and it may be difficult for the beginner to distinguish between an acidic fen and a bog on the basis of flora.

The overwhelming majority of fens in the Adirondack Mountains of New York State, most of New England, New Brunswick, Nova Scotia, and the parts of Québec covered here fall into the two more-acidic categories. Circumneutral to alkaline, or rich fens, become more abundant in parts of New York State outside the Adirondack Mountains and farther westward in our area (map 1) owing to the more widespread calcareous geological deposits there. These higher pH fens contain many herbaceous species, the majority of them of sporadic occurrence or rare, that do not occur in our acidic fens (nor in bogs). This book includes only the most common and widespread of these alkaliphiles (species that "love" alkaline conditions). You can obtain information on both the rare and common species of circumneutral-to-alkaline fens in Eggers and Reed (1997) and Kost et al. (2007), and on their identification in Chadde (2013) and Reznicek and Voss (2012).

A *bog* is a peatland with a major part of its area (along with its rooting zone) out of reach of water that has been in contact with mineral soil since it fell as rain or snow. This area of the bog is said to be *ombrotrophic*, meaning that its plants receive all their mineral nutrients from rain, snow, and dustfall. Given that these sources provide little fertility and buffering capacity, and that atmospheric inputs of acids and internal production of acids, as by *Sphagnum* or peat moss, are occurring, ombrotrophic areas are infertile and acidic (pH 3.7– 4.5).

The ombrotrophic part of a bog may coincide with a central area that is raised by peat accumulation above the level of the outer edge. Bogs with such convex surfaces are called *raised bogs*. In Maine, the raised center may be as much as 13 ft. (about 4 m) higher than the periphery, as at some coastal and near-coastal raised bogs in the eastern part of the state, or as little as 1–5 ft. (about 0.3–1.5 m) in raised bogs at the state's interior, for example, Orono Bog (map 2). As water doesn't flow uphill, water that had been in contact with mineral soil can't get to the vegetation on the raised part of such bogs, except at exceptional sites of upwelling within the peat. Other bogs are nearly flat, and distance from the bog's periphery and/or peat thickness render the bog's center ombrotrophic.

The ombrotrophic expanse of a bog may be unwooded or virtually so, and dominated by low shrubs in the heath family, along with scattered small sedges, as in some coastal Maine and New Brunswick raised bogs. Or the vegetation may be a mosaic of moss lawns (described earlier) interspersed with slightly elevated patches (hummocks) dominated by very slow-growing, dwarfed black spruce and/or American larch trees, as at most inland raised bogs of our region. Bogs in the drier climate of the upper Midwest tend to be more heavily wooded than in the maritime Northeast.

All bogs have some percentage of their area that is minerotrophic, much or all of it at the bog's periphery bordering upland, and where the peat is less deep than at the bog's center. In our region, the peripheral minerotrophic areas of bogs commonly are forested or covered by shrubby vegetation, although open graminoid areas may be present.

Fens and bogs are complex units of the landscape and often have considerable variation in soil chemistry and flora within the same fen or bog. This variation is especially prominent in bogs, with their ombrotrophic core and minerotrophic periphery.

The above concepts of bog and fen, and variations of them, developed in the early twentieth century in Scandinavia and northern Europe, have since been elaborated and refined there and elsewhere including Canada (Rydin and Jeglum 2013; Mitsch and Gosselink 2015), and continue in wide use. However, only in the past two or three decades have the concepts begun to take hold in the United States among ecologists and natural resource management personnel. In the United States, the terms "bog" and "fen" continue to be widely used in other ways. The reader may encounter these ways at some of the websites cited in the final part of the book on peatlands with boardwalks.

The most widespread use of the term "bog" by the American public is in reference to any vegetated wet place, regardless of whether peat is present, and regardless of water source, chemistry, or flora. The term "fen" barely evokes recognition by most people. Some scientists, too, have defined bog and fen differently from how I do in this book, and these alternative definitions vary widely. I will give only one example. Damman and French (1987) define a bog as "a nutrient poor, acidic peatland with a moss layer dominated by *Sphagnum* mosses; ericaceous shrubs or conifers are often dominant." This definition seems quite reasonable, on the face of it, and it correctly characterizes both chemical and floristic characteristics of ombrotrophic parts of bogs. But it ignores the complex nature of bogs and omits the fact that most bogs have minerotrophic areas, largely along their periphery but often quite extensive, that are commonly circumneutral and that support vegetation very different from that described by the authors. Further, the definition is incomplete in omitting the concept of ombrotrophy as it applies to a part of every bog, and thereby combines bogs and acidic (poor) fens with their not-so-different chemistry and flora into a single category, blurring an important distinction.

Geographic Distributions and Modes of Formation of Peatlands

The glacial history of our continent has a good deal to do with peatland abundance and distribution. An examination of a physical map of North America shows that the abundances and the areas covered by natural lakes and wetlands including peatlands are strikingly greater north of the southern limit of the most recent continental ice sheet, the Wisconsinan. This disparity, and the change in floristic character of peatlands across the glacial limit, are major reasons for my choice of that limit as the southern boundary of the area covered by this book (map 1).

Why is formerly glaciated terrain so wet? The Wisconsinan Ice Sheet disrupted the drainage of the landscape, and since the time of deglaciation the landscape has only partially recovered. The farther north toward what was the center of the former ice sheet, the shorter the period since deglaciation, and the shorter the period for recovery.

The ice flowed roughly southward in our area, reaching its limit about 18,000 years ago. As it flowed, it picked up huge numbers of rocks of all sizes, and with these rocks became a massive erosional force, grinding the landscape to widen and deepen numerous valleys. In the ensuing millennia the ice sheet melted down, and by about 10,000 years ago even its vestiges were gone from our area (ice was present later farther north). These events set the stage for the development of peatlands.

The peatlands of our glaciated terrain formed in several ways, and combinations of ways. I shall describe three ways. First, during its melting phase, the ice sheet dropped its load of rocks, more in some areas than others. These deposits blocked the drainage of numerous glacially eroded valleys to produce shallow and deep basins containing lakes. In the ensuing centuries and millennia, erosion of lake surroundings transported fine sediment, including silt, to fill parts of most lakes, and nearly all of some lakes. This infilling has been more complete in shallow basins with easily eroded catchments (watershed areas). Eventually, infilling was sufficient for wetland plants to grow and form mineral-soil wetlands. Where water flow was slow or stagnant in these wetlands, the decay of plant remains could use up oxygen in the soil, soil-water interface, and bottom water. As decay is much slower in the absence of oxygen than in its presence, further decay became slower than the addition of new plant remains to the surface. These remains continued to accumulate, transforming the mineral-soil wetland to a peatland. Over hundreds and thousands of plant generations the peat continued to deepen. As these places occupy low places in the landscape, they stay wet and continue to the present day to support living peatlands.

Second, during the ice sheet's melting phase in some areas, large blocks of glacial ice were left stranded, all or partially buried in deep glacial deposits. These blocks subsequently melted, to produce basins called ice-block depressions, kettle holes, or simply *kettles*. These basins filled with water up to the depth of their surrounding water tables, to form seepage lakes or ponds. Such water bodies exchange water largely by slow subsurface inflows and outflows rather than permanent surface inlet and outlet streams. Over the millennia, these lakes and ponds were invaded by wetland plants. Under conditions of slow water exchange, shelter from the wind, and little water circulation, oxygen was easily depleted, allowing plant remains to accumulate as peat. Typically, this accumulation progressed from the basin edge inward, and in many kettle peatlands open water still remains as a central pond. In the southern parts of our area (map 1), a large percentage of the bogs and fens occupy kettles, as I indicate for many of the peatlands with boardwalks that I list in the final part of this book. Peatlands in kettles also occur in the northern parts, but the abundance of non-kettle peatlands greatly increases northward, and most of these are much larger than those in kettles.

The third way is the initiation of peat accumulation on a soil surface wet enough for development of anoxia in the accumulating plant remains, but with no fully aquatic phase. This process is called *paludification*. The term, as used here, includes peatland initiation on freshly formed surfaces ("primary mire formation"), as on recently deglaciated terrain, as well as peatland initiation in already existent terrestrial ecosystems, as in cleared upland forests. The latter form of paludification is the way the blanket bogs got started on sloping terrain in Ireland, with its cool and wet maritime climate. It is perhaps surprising that some of our peatlands also got started that way. In Maine, for example, where the deepest peat and underlying deposits from multiple locations in 96 of the state's peatlands have been sampled, aquatic deposits were absent from several of the peatlands, leading to the conclusion that these several peatlands started by paludification (Sorenson 1986; peatland descriptions in Davis and Anderson 1999). Additionally, in some of the state's raised bogs with an initial aquatic phase, later lateral spread of the peatland was by paludification (Perkins 1985, Sorenson 1986, Davis and Anderson 1991 and 1999, and Hu and Davis 1995). North of our area, paludification has been an important process in the formation and spread of peatlands (Kuhry and Turunen 2006).

The astute reader will already have realized that both local and regional factors are involved in the formation and persistence of peatlands. Local factors such as basin drainage, permeability to water of underlying mineral deposits, and runoff from adjacent uplands are all important. On a broad regional scale, the abundance of peatlands is greatly under the influence of climate. On this scale, some minimal excess of precipitation over evaporation is needed. Evaporation, largely a function of temperature, decreases northward in our hemisphere. Even in the relatively small region covered by this book, peatland abundance and total area generally increase northward. The existence of these soggy landscapes results from a complex balance of many factors.

Peatlands and Peatland Plants beyond Our Geographic Area

Bogs and fens occur in many parts of the world, wherever landscape hydrology and climate permit, even including some wet areas of the tropics. Examples include the pocosin peatlands of the Carolinas; major parts of the Florida Everglades; the peat swamp forests of Java, Sumatra, and Borneo; the wooded and grassy peatlands of the New Guinea highlands; the peatlands of Tasmania; and the vast peatlands of central and northern Eurasia. The vascular plant floras of peatlands remote from our geographic area substantially differ from the flora I cover in this book, except that Northern Hemisphere boreal peatlands around the world share many species whose distributions are *circumboreal*. A great east-west band, albeit a discontinuous swath, of peatland spreads across the boreal and subarctic regions of the Northern Hemisphere. A naturalist familiar with the peatlands of the northern parts of the region covered by this book would feel quite at home in the peatlands of Finland and Siberia.

The boreal flora of this soggy swath peters out at its southern fringe. We are at the southern fringe. Many of the species of our peatlands, including some that are abundant members of the vegetation, reach their southern continental limit in our area. Prominent examples include black spruce, Labrador tea, and tall cottonsedge. Their abundances are greatest in the northern parts of our area, and their distributions extend far north of the 49° N latitudinal limit of the area (map 1). These and other northern species give a decidedly boreal character to most of our bogs and fens.

Further Reading: An Annotated List

This annotated list includes several general peatland references that apply well to all 19 states and provinces covered by the book. Some other references pertain directly to Maine peatlands, the state I am most familiar with. These Maine references are more or less useful in adjacent states and provinces, as noted. I counterbalance the Maine emphasis with a few references on the western part of our area where circumneutral and alkaline fens are more numerous, as in the states of Michigan and Wisconsin. Not included here is peatland literature applying specifically to each of the other 16 states and provinces. For the reader seeking greater detail on the peatlands of one of these states or provinces, a good place to start is at the website of the jurisdiction's natural resource agency or ministry, and by seeking the web pages on the state's natural areas and plant communities.

Anderson, D. S., and R. B. Davis. 1997. "The Vegetation and Its Environments in Maine Peatlands." *Canadian Journal of Botany* 75:1785–1805. *The peatland vegetation and physical and chemical environments of Maine peatlands resemble those in nearby jurisdictions and are more or less well represented in peatlands throughout our region. However, as the vast majority of Maine peatlands are acidic, this reference is not as useful in circumneutral and alkaline peatlands, which are more abundant in some parts of other states and provinces. The same proviso applies to the other Maine references that follow.*

———. 1998. *The Flora and Plant Communities of Maine Peatlands.* Maine Agricultural and Forestry Experiment Station, Orono. Technical Bulletin 170. *The list of peatland species in this paper is much longer than the one in this book.*

Anderson. D. S., R. B. Davis, and J. A. Janssens. 1995. "Relationships of Bryophytes and Lichens to Environmental Gradients in Maine Peatlands." *Vegetatio* 120:147–59. *This paper lists the peat mosses and other bryophytes as well as terrestrial lichens of Maine peatlands, and the physical and chemical factors correlated with their occurrence.*

Anderson, D. S., R. B. Davis, S. C. Rooney, and C. S. Campbell. 1996. "The Ecology of Sedges (Cyperaceae) in Maine Peatlands." *Torrey Botanical Club Bulletin* 123:100–110. *This reference lists the many sedge species in Maine peatlands and the physical and chemical factors correlated with their occurrence.*

Arsenault, M., G. H. Mittelhauser, D. Cameron, A. C. Dibble, A. Haines, and J. E. Weber. 2013. *Sedges of Maine: A Field Guide to Cyperaceae.* Orono: University of Maine Press. *Although ostensibly covering only Maine, this book includes all the sedge species common in the peatlands of our area. Because of its extensive use of photographs, the book may be the most accessible identification manual to the sedges of our peatlands.*

Chadde, S. W. 2013. *Wisconsin Flora: An Illustrated Guide to the Vascular Plants of Wisconsin.* CreateSpace Independent Publishing Platform. *This is the most up-to-date manual on the seed plants of Wisconsin and is applicable to adjacent states. It includes species of northern Midwestern peatlands that do not appear in Haines (2011).*

Crum, H. 1992. *A Focus on Peatlands and Peat Mosses.* Ann Arbor: University of Michigan Press. *This book is a semipopular summary of what was known about peatlands and peat mosses, with emphasis on north-Midwestern United States.*

Damman, A. W. H., and T. W. French. 1987. *The Ecology of Peat Bogs of the Glaciated Northeastern United States: A Community Profile.* Biological Report 85(7.16), U.S. Fish and Wildlife Service, Washington, DC. *This report covers major aspects of the ecology of the bogs and fens of northeastern United States.*

Davis, R. B., and D. S. Anderson. 1991. *The Eccentric Bogs of Maine: A Rare Wetland Type in the United States.* Maine Agricultural Experiment Station, Orono. Technical Bulletin 146. *Based on a hydrogeomorphic classification of peatlands, this book singles out a unique type that is rare in the entire area covered by this book.*

———. 1999. *A Numerical Method and Supporting Database for Evaluation of Maine Peatlands as Candidate Natural Areas.* Maine Agricultural and Forestry Experiment Station, Orono. Technical Bulletin 175. *This publication contains extensive biological and physical data including vegetation cover maps of a representative set of 102 of Maine's diverse peat-*

lands, most of them similar to peatlands in adjacent states and provinces.

———. 2001. "Classification and Distribution of Freshwater Peatlands in Maine." *Northeastern Naturalist* 8:1–50. *This long article presents a hydrogeomorphic classification of peatlands as units of the landscape, supporting data, geographic distributions of peatland types, and a color aerial photograph of each type. The conceptual framework of classification is relevant in neighboring states and provinces.*

Eggers, S. D., and D. M. Reed. 1997. *Wetland Plants and Plant Communities of Minnesota and Wisconsin.* 2nd ed. St. Paul, MN: U.S. Army Corps of Engineers. *The authors describe the plant communities and their environments in the peatlands and other wetlands of our northwesternmost state (Wisconsin). The book is of especial interest to readers of this book as a supplementary source of information on the flora of the alkaline fens of the north-Midwestern states covered by this book.*

Gleason, H. A. 1968. *The New Britton and Brown Illustrated Flora of the Northeastern United States and Adjacent Canada.* Vols. 1–3. New York: New York Botanical Garden. *Although some of the Latin names have changed since this classic's publication, it remains the only nearly complete illustrated (with line drawings) flora of the entire region covered in this field guide.*

Haines, A. 2011. *New England Wild Flower Society's Flora Novae Angliae: A Manual for the Identification of Native and Naturalized Higher Vascular Plants of New England.* New Haven, CT: Yale University Press. *This flora is the most up-to-date and complete for New England and presents taxonomic keys that include all the species in this field guide, plus brief descriptions of each one. Its species names are followed here.*

Hu, F. S., and R. B. Davis. 1995. "Postglacial Development of a Maine Bog and Paleoenvironmental Implications." *Canadian Journal of Botany* 73:638–49. *This paper provides readers with an example of a paleoecological reconstruction, based on multiple peat cores, of the development of a raised bog.*

Johnson, C. W. 1985. *Bogs of the Northeast.* Hanover, NH: University Press of New England. *This popular book on the peatlands (fens included, despite only "bogs" in the title) of the northeastern United States includes line drawings of some common plants and animals.*

Keddy, P. A. 2010. *Wetland Ecology: Principles and Conservation.* 2nd ed. Cambridge: Cambridge University Press. 516 pp. *This is a leading university textbook covering all aspects of wetlands, including peatlands.*

Kost, M. A., D. A. Albert, J. G. Cohen, B. S. Slaughter, R. K. Schillo, C. R. Weber, and K. A. Chapman. 2007. *Natural Communities of Michigan: Classification and Description.* Lansing: Michigan Natural Features Inventory, Report No. 2007–21. *This book does for the entire state of Michigan, including its peatlands, what Eggers and Reed (1997) do for Wisconsin wetlands, but also includes upland plant communities.*

Kuhry, P., and J. Turunen. 2006. "The Postglacial Development of Boreal and Subarctic Peatlands." In *Boreal Peatland Ecosystems (Ecological Studies),* edited by R. K. Wieder and D. H. Vitt, 25–46. Berlin: Springer. *This book chapter describes the modes of peatland initiation and subsequent patterns of development, and the external and internal forces driving these patterns.*

Martini, I. P., A. M. Cortizas, and W. Chesworth, eds. 2007. *Peatlands.* Vol. 9 of *Evolution and Records of Environmental and Climatic Changes (Developments in Earth Surface Processes).* Amsterdam: Elsevier Science. *The volume is a comprehensive scientific treatment of the world's peatlands, including their geological "evolution," ecology, chemistry, paleoecology, peat resources, human impacts, climate change impacts, and other.*

McQueen, C. B. 1990. *Field Guide to the Peat Mosses of Boreal North America.* Hanover, NH: University Press of New England. *This book is a guide for field identification (with a strong hand lens) of the common peat moss species of our peatlands. It is suitable for an amateur naturalist.*

Mitsch, W. J., and J. G. Gosselink. 2015. *Wetlands.* 5th ed. New York: John Wiley & Sons. 744 pp. *This is a leading university textbook that covers all major aspects of wetlands, including peatlands.*

Niering, W. A. 1985. *Wetlands.* Audubon Society Nature Guides. New York: Knopf. *This book is a comprehensive illustrated field guide to the wetlands of the United States, including both plants and animals. The book's extensive coverage precludes intensive treatment of northern peatlands.*

Perkins, J. 1985. "A Palaeoecological Study of the Smith Brook

Deadwater Peatland, Maine, U.S.A." University of Sheffield, UK, honors thesis. *This study, based on multiple peat cores from a raised, eccentric bog, demonstrates aquatic origins of most of the peatland, plus overtopping of rock ridges and limited spread to adjacent uplands by paludification.*

Reznicek, A. A., and E. G. Voss. 2012. *Field Manual of Michigan Flora.* Ann Arbor: University of Michigan Press. *This is the most up-to-date and complete manual to the seed plants of Michigan and is largely applicable to adjacent states and the area of Canada just to the north. It includes peatland species of northern Midwestern peatlands that do not appear in Haines (2011).*

Rydin, H., and J. K. Jeglum. 2013. *The Biology of Peatlands.* Biology of Habitats. 2nd ed. Oxford: Oxford University Press. *This is a leading university textbook covering all aspects of peatlands.*

Schnell, D. E. 2002. *Carnivorous Plants of the United States and Canada.* 2nd ed. Portland, OR: Timber Press. 470 pp. *A comprehensive description of the carnivorous plants, including all the species that occur in the peatlands covered by this book.*

Sorenson, E. R. 1986. *Ecology and Distribution of Ribbed Fens in Maine and Their Relevance to the Critical Areas Program.* Planning Report 81, Maine State Planning Office, Augusta.

Plant Structure Terminology Used in the Species Descriptions

An understanding of the species descriptions depends on familiarity with some basic aspects of plant structure. For each species in the book, I describe both vegetative (stems, leaves, roots, etc.) and reproductive (flowers, fruits, cones, etc.) features using a minimum amount of technical jargon, but necessarily must include some botanical terms. As in prior parts of the book, I draw attention to important terms by italicizing them. For most of the terms, I give examples from among the species descriptions. The reader may go to those descriptions to see photographs of the structures. Readers already familiar with the basics of plant structure may wish to skip the rest of this section.

Flowering Plants

I start with flowering plants, and define and describe the nonreproductive and reproductive structures that are most useful for identifying them to species.

NONREPRODUCTIVE OR VEGETATIVE STRUCTURES

The stems of flowering plants often have distinctive features that are useful for species identification, including the color of small stems (e.g., red-osier dogwood), and several characteristics of the bark of large stems, for example, the peeling bark of highbush blueberry. Another useful feature of stems for identification is the color of the *pith* or soft center of the stem (e.g., red-osier dogwood). The *nodes* are the places along stems where leaves, buds, and branches are attached.

Leaves may be simple or compound. A *simple leaf* of a broadleaved plant typically consists of two parts: the flat leaf *blade* and the narrow *petiole*. The blade may have veins that extend more or less straight from the midrib to the leaf margin, or the veins may be *arcuate*, meaning that they curve in an arc more or less parallel to the leaf edge (e.g., alder-leaved buckthorn). Some of our species have leaf blades that are *revolute*, meaning that their edges are rolled under, and others are *involute*, meaning that their edges are rolled upward. The petiole attaches the leaf blade to the stem (e.g., red maple; mountain holly). Some species have simple leaves that lack a distinct petiole, and the blade appears to arise directly from the stem (e.g., sweetgale). Before describing leaves that are compound (not simple), it is necessary to discuss the buds.

Buds contain miniature stems, leaves, and/or flowers, and in woody plants they overwinter in a dormant condition. In most woody plants, buds are protected by one or more outer *bud scales*, but in some species buds lack scales and are said to be *naked buds* (e.g., withe rod). When *buds* arise from the stem in the *axils* of leaves—that is, in the "armpit" where the petiole or blade joins the stem—they are called *axillary* buds (e.g., black ash; mountain honeysuckle). A bud that is borne at the tip of the stem is called a *terminal* bud (e.g., black ash; mountain honeysuckle). In the spring, the buds of woody plants, and of herbaceous plants with rhizomes, expand and open to form new stems, leaves, and/or flowers.

In *compound leaves*, the blade is subdivided into multiple *leaflets* (e.g., black ash). The leaflets all attach to a *rachis*. The rachis is the equivalent of the midrib of a simple leaf. It extends to the stem in same fashion as the midrib and petiole of a simple leaf. While leaflets may resemble full leaves to the unaware observer, the point at which a leaflet attaches to the rachis bears no bud. In woody and many herbaceous plants with compound leaves, an axillary bud is present on the stem where the rachis of the leaf attaches to it. In such plants, look for the presence of a bud to determine if you are observing a leaflet (no bud in its axil) or a leaf (a bud in its axil).

Leaves may arise from a stem node in pairs, each member of a pair directly across the stem from the other, an *opposite-leaved* arrangement (e.g., red-osier dogwood; American water-horehound), in multiples greater than two, a *whorled* leaf arrangement (e.g., sheep laurel), or they may arise individually, each leaf on the opposite side but farther along the stem from the prior one, an *alternate-leaved* arrangement (e.g., gray birch).

REPRODUCTIVE STRUCTURES

The four major parts of the flower are, starting at the flower's attachment to the stem: (1) a whorl of *sepals* (typically green), (2) a whorl of *petals* (any color, and often showy), (3) a whorl of *stamens* ("male" part), and (4) one or more *pistils* ("female" part). The stamens each consist of a filament bearing an *anther* at the tip. The anther produces the pollen. A typical pistil consists of a basal *ovary*, and a relatively narrow *style* extending up from it. The style has a *stigma* (pollen-receptive surface) at the tip. After pollination, one or more seeds develop in the ovary, and the pistil develops into a *fruit*. For diagrams of flower structures and their variations go to http://en.wikipedia.org/wiki/Flower.

In most flowering plant species, each flower contains both stamens and one or more pistils. Flowers with both these parts are called *perfect*, but in some species flowers are *imperfect*. Imperfect flowers with only stamens are called *staminate* flowers, and those with only pistils are called *pistillate* (or *carpellate*) flowers. In some species, separate staminate and pistillate flowers occur on the same plant, as in gray birch and speckled alder, but in other species they are confined to separate plants, as in red maple, common winterberry, and sweetgale.

Flowers may be borne individually on a separate stalk (e.g., grass-of-Parnassus), individually among the leaves (e.g., common winterberry staminate plant; large cranberry), or on a special stem where they form a cluster of flowers. Such a flower cluster is called an *inflorescence* (e.g., arrowwood; swamp candles). The positions on the plant, shapes, and other aspects of inflorescences vary from one species to another. The variations most relevant to identification of the species featured in this book are described in the following paragraphs.

A *spike* is a typically narrow and elongate inflorescence bearing flowers that lack stalks. Spikes and diminutive spikes called *spikelets* characterize the family with the most species in our peatlands, the Cyperaceae (sedges). The grasses (Poaceae) also have spikelets. A variation of a spike is a *spadix* or fleshy spike. It occurs in skunk cabbage and wild calla, where it is subtended by a large bract called a *spathe*.

A special type of inflorescence called a *head* occurs in the composite family, Asteraceae. Several species in that family are described in this book (e.g., calico aster). A head is composed of a compact assemblage of flowers. Around the outside of the head in some species, *ray flowers* with more or less flat rays are present (visualize the white rays of a daisy). They may be fertile or just to attract insect pollinators, depending on the species. Ray flowers are sometimes incorrectly called petals. All heads have a compact disk or hemisphere consisting of many tightly packed *disk flowers* (visualize the yellow center of a daisy). Unlike daisies and asters, many members of the family have heads without ray flowers (e.g., devil's beggar-ticks). The older name for the Asteraceae is Compositae, because what appears to be one flower is actually a *composite* of smaller flowers. The species of the family are collectively referred to as "composites." The term "head" is also applied to compact inflorescences in other families.

Another type of inflorescence is the *catkin* or ament. Many tiny flowers are packed together on an elongate axis, each flower separated from the next one by a bract. Some catkins hang down when mature, unlike more or less upright, bractless spikes. Catkins occur in the birch family, Betulaceae, and in other families with representatives in peatlands.

For much of the year when we visit peatlands, the plants are in fruit rather than in flower, so it is helpful for plant identification to know what their fruits look like. A unique aspect of this plant guide is its near-consistent inclusion of a description and/or photo of the fruit and the naming of the fruit type of each species. Most species in this book have *simple fruits*. Each of these is derived from a flower with a single pistil, and is either dry or fleshy.

Dry simple fruits lack a fleshy part, are produced by many peatland plants, and include the following types. An *achene* is derived from a simple pistil, contains a single seed, and has a hard outer case that lacks a suture to split open at maturity. It occurs in sedge species and numerous other species featured in the book. A *sa-*

mara is a winged achene. It occurs in ashes, maples, and birches. A *capsule* is derived from a compound pistil formed of two or more joined pistils, each forming a separate chamber (locule) of the fruit. Most capsules have chambers with sutures that split open at maturity to release seeds. Capsules are produced by the members of the heath family and other families with species in this book. A *follicle* is formed from a simple pistil, contains two or more seeds, and has a single suture that splits open at maturity to release seeds. It occurs in meadowsweet species, podgrass, and other plants described in this book. A *nut* or *nutlet* is formed from a single, simple ovary, has a single seed, and a thick, hard outer wall that lacks a suture for splitting open, as in sweetgale.

Fleshy simple fruits are also commonly produced by peatland species and include berries and drupes. A *berry* is derived from one flower with one ovary whose wall ripens into typically edible soft tissue. It is produced by blueberry and cranberry species. But many other peatland plants with "berry" in their name, like winterberry, huckleberry, and blackberry, do not produce berries, but produce drupes or other fleshy fruits. Some, like maleberry, produce a dry fruit, a capsule in its case. A *drupe* is a simple fleshy fruit derived from a single pistil that has a hard stony layer surrounding the seed (think peach). Drupes are produced by arrowwood, poison-sumac, withe rod, baked-apple berry, and many others.

Not all peatland species produce simple fruits. Some produce *aggregate fruits*. These are produced from a single flower with multiple and separate simple pistils. Each pistil produces a simple fruitlet that can be a berry, drupe, or other. Peatland examples include bristly blackberry and baked-apple berry, each of which has a "berry" that is really an aggregate fruit made up of many small drupes. Other peatland species have *accessory fruits*. In these, the fleshy outer part is not derived from the pistil or pistils but has grown around it or them. Peatland examples include black chokeberry (a *pome*, like an apple) and shining rose (a *hip*). Skunk cabbage produces a *multiple fruit*, a combination of the fruits from multiple, separate small flowers in an inflorescence. The fruits from these flowers are embedded in the spadix (see skunk cabbage description). More information on fruit types can be found at http://en.wikipedia.org/wiki/Fruit.

Conifers

The needles (e.g., black spruce) or scales (e.g., northern white cedar) are actually leaves. Conifers do not have flowers or fruits but have cones (strobili) that produce pollen or seeds. The cones are of two separate types, *pollen-producing cones* and *seed cones*. The pollen is wind-dispersed.

Species Descriptions

Some features that distinguish one species from another are tiny. When you venture into the field, you will find it helpful to bring a hand lens or strong magnifying glass. If you are a birder, you may use one side of your binoculars in reverse direction as a weak microscope. Look in the wide end, and move the eyepiece end to about half an inch (about 1 cm) from the subject.

The following species descriptions are grouped by growth form. These groups appear in the same order as they were described in the first part of the book, starting with trees. The edge of each page is color coded to indicate the growth form of the plant. Within each growth form, species are listed in alphabetical order by common name. However, when a common name consists of more than one word, and one of the words is the equivalent of the Latin genus, I use that word for the primary step in alphabetization. For example, the species black spruce is alphabetized under *s*, not *b*, as spruce is the common name for the genus of all spruce species. Similarly, if you see a shrub that you suspect is highbush blueberry, find the description under *b*, not *h*. Notable exceptions include the grasses and orchids, and most sedge species. All the grass species are grouped together under *g*, all the orchid species under *o*, and sedge species with "sedge" in their common name are grouped under *s*.

In the descriptions, the numbered photos are referenced by numbers in parentheses. References to "our area," "our region," "our states and provinces," or our "jurisdictions" refer only to the area covered by this book (map 1), unless otherwise noted.

For each species, an abbreviation in capital letters is used to indicate the degree to which the species is confined to wetlands or, conversely, to uplands (non-wetlands). The words represented by these abbreviations, in order from almost always found in wetlands to almost always found at uplands, are: OBL = obligate wetland; FACW = facultative wetland; FAC = facultative; FACU = facultative upland; and UPL = obligate upland. These terms were explained in the section Wetland Indicator Status of Plant Species.

1

Black ash (*Fraxinus nigra*) is a midsize deciduous tree in the olive family, Oleaceae. It is also called basket ash because it is used by the Native Americans of our region for making baskets.

IDENTIFICATION The opposite leaves are compound and arise from stout twigs. The leaves may have as many as 15 leaflets but commonly have 5 to 11 (1.1). Other ash species have the same basic leaf structure and arrangement and are easily confused with each other. One of these other species in our region, green ash (*F. pennsylvanica*) regularly occurs (FACW) in wetlands, but typically not on peat soils. Black ash is distinguished by leaflets that arise directly from the rachis, whereas leaflets of the other two common species of our region, green and white ash (*F. americana*—FACU) have short stalks that attach them to the rachis. Additionally, black ash has a blob of tomentum (like cotton) where each leaflet attaches to the rachis (1.2), and the terminal bud and first pair of axillary buds are separate from each other (1.3). In the other two species these buds abut each other. Black ash flowers are small, lack petals, and occur in inflorescences (1.4) that emerge in spring from axillary buds just before emergence of the current season's leaves. The winged fruits (samaras) are 1.0–1.5 in. (2.5–4.0 cm) long (1.5).

OCCURRENCE · FACW Black ash occurs in all our states and provinces and is found at wooded and sparsely wooded sites in fens, at the wooded minerotrophic periphery of bogs, and at swamps and moist woods throughout the area.

1.4

1.1 Black ash: compound leaves. The leaf filling the upper right two-thirds of the photo has seven leaflets. 1.2 Black ash: leaf detail. White arrow indicates blob of cottony tomentum at the base of a leaflet. 1.3 Black ash: buds. Arrow 1 indicates a terminal bud, and arrow 2 an axillary bud. 1. 4 Black ash: staminate inflorescences. Photograph: Marilylle Soveran. 1. 5 Black ash: winged fruits (samaras). Photograph: Anton A. Reznicek.

2

Gray birch *(Betula populifolia)* is a small deciduous tree in the birch family, Betulaceae.

IDENTIFICATION The bark is dirty white and blotched with gray or black (2.1) and does not peel in large sheets like paper birch, *B. papyrifera* (FACU). The triangular leaves are alternate. They may appear opposite where they arise very close to each other on slow-growing spur branches, but careful observation confirms their alternate arrangement (2.2). Many small flowers are packed together in elongate catkins. When mature, staminate catkins are flaccid and hang down (2.3). Pistillate catkins are stiff and may point in various directions when they are receptive to pollen (2.4). Mature pistillate catkins contain numerous small winged fruits (samaras). Typically the catkins shed their fruits in winter and disintegrate, but some persist to the next spring (2.4). Commonly the fruits are wind-dispersed in winter over a snow surface.

OCCURRENCE · FAC This species is found in parts of all our jurisdictions except Wisconsin. It is most abundant at wooded sites that have been logged or otherwise disturbed in the prior few decades. It grows in both wetlands and non-wetlands, including abandoned fields. In peatlands it occurs at wooded fens and the minerotrophic periphery of bogs.

2.2

2.1 Gray birch: tree trunk showing white bark mottled with gray and black.
2.2 Gray birch: branches with alternate leaves. **2.3** Gray birch: a hanging staminate catkin at pollen-shedding stage, and upright pistillate catkins (green); mid-May in central Maine. **2.4** Gray birch: a young, upright pistillate catkin (green) at the pollen-receptive stage, and two mature hanging pistillate catkins (gray brown) that were formed the prior year; mid-May in central Maine.

3

Northern white cedar (*Thuja occidentalis*) is a small to medium-size, often irregularly shaped evergreen tree in the Cupressaceae or cypress family. It also goes by the name arborvitae. The Atlantic white cedar (*Chamaecyparis thyoides*) also occurs in our area.

IDENTIFICATION Northern white cedar branchlets are flat with scalelike leaves (3.1). The pollen-producing cones (3.2) form at the tips of branchlets, are only about 0.08 in. (approx. 2 mm) across, and shed their pollen in the spring. The seed cones (3.3) are elliptical, and green when young in the spring. By the fall when they start shedding their winged seeds they have become tan and woody (3.4) and about 0.4–0.5 in. (1.0–1.3 cm) long. By contrast, Atlantic white cedar has branchlets that are round or squarish in cross section, and has nearly spherical seed cones.

OCCURRENCE · FACW Northern white cedar occurs in the less acidic parts of our peatlands in all our states and provinces, commonly at wooded and open-wooded parts of fens, and at the wooded periphery of bogs. It becomes rare in the southern part of our area. The species also occurs at uplands, often at moist spots or seeps. Atlantic white cedar (OBL) occurs in wooded fens and swamps along the Eastern Seaboard from southern Maine to northern Florida and along the eastern Gulf Coast, as far as 100 mi. inland (about 160 km) at a few sites, but largely within 50 mi. (about 80 km) of the coast.

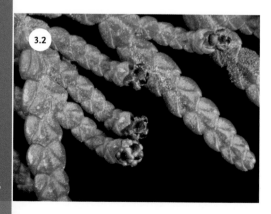

3.2

3.1 Northern white cedar: branch showing scalelike leaves. **3.2** Northern white cedar: branchlets with pollen-producing cones at tips. Photograph: Graham R. Powell. **3.3** Northern white cedar: branch with seed cones in midsummer at central Maine. **3.4** Northern white cedar: branch with seed cones in May in central Maine. These cones had shed their seeds the prior fall and winter.

4 Balsam fir

Balsam fir (*Abies balsamea*) is a medium-size conical evergreen tree in the family Pinaceae. It is commonly used as a Christmas tree, as it retains its needles long after it is cut, and it is grown commercially for this purpose.

IDENTIFICATION This species grows to about 55 ft. (about 18 m) tall. The flat needles arise individually around the stem but are oriented in the same plane when in the shade (4.1, 4.2). Balsam fir seed cones are 2–3 in. (5–7.5 cm) long and stand up like candles atop the tree (4.3). They dry out and turn brown over the summer and shed their winged seeds in the fall. The pollen-producing cones are about 0.2 in. (about 0.5 cm) long and are short-lived after shedding their pollen in the spring (4.4).

In our region balsam fir can be confused with eastern hemlock (*Tsuga canadensis*—FACU), an evergreen tree that also has flat needles. Each hemlock needle has a short, fine stalk connecting it to the stem, whereas each balsam fir needle is attached by a short stump (4.2). Eastern hemlock seed cones are small—about 1 in. (2.5 cm)—and hang down. The only other species with flat needles that might be might be confused with balsam fir is the evergreen shrub American yew (*Taxus canadensis*—FACU). Its needles are tipped by a sharp point and have yellowish undersides. It produces its seeds singly in a fleshy covering (red at maturity) instead of in a cone. It typically is not taller than 4 ft. (1.2 m), exceptionally to 6 ft. (1.8 m).

OCCURRENCE · FAC Balsam fir occurs in all our states and provinces except Illinois. It grows in wooded fens and on hummocks at the minerotrophic wooded outer parts of bogs. This species also occurs extensively in upland forests, most abundantly in the northern half of our region. In the southern half, it is largely confined to mountains and cool moist glens. Eastern hemlock also occurs in all our states and provinces except Illinois. It grows in wooded fens in the southern half of our area and extensively at moist uplands. In the eastern part, it often occurs in wooded fens with Atlantic white cedar and black gum or tupelo (*Nyssa sylvatica*—FAC).

4.1 Balsam fir: branches showing flat needles. Photo taken in early July in central Maine when the new needles were still a lighter shade of green. **4.2** Balsam fir: close-up of needles. Photograph: Glen H. Mittelhauser. **4.3** Balsam fir: erect seed cones. **4.4** Balsam fir: pollen-producing cones that had recently shed their pollen. Photograph: Graham R. Powell.

5

American larch (*Larix laricina*), in the family Pinaceae, is a deciduous conifer that locally goes by the name tamarack or hackmatack.

IDENTIFICATION Larch is an irregularly conical tree (5.1) that turns "smoky gold" in the fall (5.2) before it drops its needles. A full-size tree can be 75 ft. (23 m) tall, but growth is severely stunted at the ombrotrophic center of our bogs (5.1). Branches in leaf may have a feathery appearance. The needles occur in tufts of 10 to 30 (rarely more, but sometimes as many as 60) and are 0.6–1.0 in. (1.5–2.5 cm) long (5.3). The pollen-producing cones (5.4) are 0.4–0.5 in. (1.0–1.3 cm) long and are short-lived after shedding their pollen in the spring. The seed cones (5.3) are 0.5–0.8 in. (1.3–2.0 cm) long and shed winged seeds in the fall.

OCCURRENCE · FACW American larch may occur at any part of a bog or fen, except in open water. It generally is the second most abundant coniferous tree in bogs, after black spruce, but at western parts of our area it can be more abundant than black spruce in some peatlands. The species occurs in all the states and provinces of our area but is more abundant in the northern part of the region, where it also occurs outside of peatlands on moist mineral soils.

5.3

5.1 American larch: this slow-growing specimen about 10 ft. (approx. 3 m) tall was growing near the center of a bog. **5.2** American larch: in the fall before dropping needles. **5.3** American larch: branch with needle bunches and seed cones.
5.4 American larch: branch with pollen-producing cones and emerging needle clusters in spring. Photograph: Glen H. Mittelhauser.

6

Red maple

This deciduous tree species (*Acer rubrum*) is also called swamp maple. It is in the family Sapindaceae.

IDENTIFICATION The leaves are opposite. The blades are five-lobed with the basal pair of lobes the smallest. The lobe margins are irregularly serrated (6.1). Leaf undersides are whitened (6.2). The petioles are red during the entire season (6.1). In the fall the blades commonly turn red (6.2), too, but may turn orange or yellow instead. The staminate (6.3) and pistillate (6.4) flowers bloom on separate trees in the spring prior to leaf emergence. The fruits are samaras that are attached in pairs (6.5). A closely related tree species, silver maple (*A. saccharinum*—FACW), has leaves with deeper notches between the lobes and occurs mainly in mineral-soil wetlands.

OCCURRENCE · FAC This widespread species occurs in wetlands and non-wetlands throughout our area and beyond. It is commonly the dominant deciduous flowering tree species in wooded fens and the peripheral, minerotrophic areas of bogs but is less common in the far north of our area.

6.1

6.1 Red maple: lobed leaves. **6.2** Red maple: leaves in the fall. Some of the leaves show their whitened (appear pink) undersides. **6.3** Red maple: staminate flowers. Photograph: Glen H. Mittelhauser. **6.4** Red maple: pistillate flowers. Photograph: Glen H. Mittelhauser. **6.5** Red maple: winged fruits (samaras). Photograph: Glen H. Mittelhauser.

7

Black spruce (*Picea mariana*) is an evergreen coniferous tree in the family Pinaceae. Barring fire at open and open-wooded areas of bogs and acidic (sphagnous) fens, reproduction of black spruce by seed is rare. At such areas, its primary means of reproduction and spread is by a form of vegetative reproduction called *layering* (7.1). The few tallest trees in the foreground of photo 7.1, each 5–8 ft. (1.5–2.4 m) tall, very slow-growing, and 70–110 years old, are ringed by trees only 1–2 ft. (0.3–0.6 m) tall. A few decades before the photo was taken, the lowest branches of the taller trees were overgrown by peat moss, grew upward at their tips, and took root to form the ring of small trees. This entire group of trees is a clone, with all individuals being genetically identical. Eventually, the largest and oldest central tree(s) will die, and another tree or small number of trees from the ring will become central to a layering group. Potentially, and barring fire, this replacement process will change the clone's position. These clones may be very old, but no one knows how old.

IDENTIFICATION Typically, black spruce has a slender form (7.2) and rarely grows taller than 75 ft. (23 m). At ombrotrophic sites in bogs it grows very slowly and is dwarfed. At such sites, an 8-ft.-tall (2.4 m) tree may be more than 100 years old (7.1). The branches

7.1

7.1 Black spruce: dwarfed trees reproducing by layering at ombrotrophic area of a bog. **7.2** Black spruce: slender trees at the minerotrophic-ombrotrophic transitional zone of a bog.

7.3 Black spruce: needles arising from a hairy stem. Photograph: Glen H. Mittelhauser. **7.4** Black spruce: pollen-producing cones shortly before reaching pollen-shedding stage. Photograph: Graham R. Powell. **7.5** Black spruce: seed cones persist for many years on the old parts of branches.

have short (about 0.6 in. or 1.5 cm) needles with a squarish cross section, and these needles arise all around a hairy stem (7.3). The needles are commonly gray green but in some instances are yellow green. The pollen-producing cones (7.4) are formed on the outer branches of the tree crown below the current year's seed cones. They are 0.4–0.7 in. (1.0–1.8 cm) long, and red until shedding their pollen. After shedding, they fall off the tree. The seed cones (7.5) are 0.6–1.0 in. (1.5–2.5 cm) long, exceptionally to 1.3 in. (3.3 cm) and are retained on the tree for many years, as shown by their positions close to the trunk on the old parts of branches (7.5). These cones usually remain closed on the tree with seed dissemination delayed. Fire commonly is the cause of release of the winged seeds.

The closely related red spruce (*P. rubens*—FACU) generally grows taller than black spruce and has a broader shape. Its needles are more consistently yellow green, and it has longer (1.2–1.6 in.; 3.0–4.1 cm) seed cones. These cones aren't persistent on the tree like those of black spruce. They shed their winged seeds when they first mature and typically fall from the tree within a year after shedding. Hybrids of these two species, with intermediate features, occur in some areas.

OCCURRENCE · FACW Black spruce occurs in all states and provinces of our area except Ohio and Indiana. It typically is the most abundant and dominant conifer in bogs and acidic fens of the eastern, central, and northern parts of the region. It also occurs at non-wetlands: high on the northeastern mountains and at some coastal sites from eastern Maine northward. It is very restricted in occurrence in the southern parts of our area. Red spruce occurs in forested peatlands at the southeastern parts of our area and more widely in uplands of the Northeast as far north as southern Québec and southeastern Ontario.

8

Speckled alder This tall shrub, *Alnus incana* subspecies *rugosa*, is a member of the birch family, Betulaceae. It has nitrogen-fixing bacteria in root nodules that provide fixed nitrogen to the host plant, soil, and other plants.

IDENTIFICATION Speckled alder can grow 25 ft. (8 m) tall, but it is shorter than 15 ft. (4.5 m) in our peatlands. It has alternate leaves with fine serrations on a wavy or coarsely toothed edge (8.1). The buds are stalked (8.2). Small staminate and pistillate flowers form in separate staminate and pistillate catkins, respectively, on the same plant. The formation of reproductive structures and the dispersal of fruits is a two-year process in this species. In the first growing season the catkins form but stay closed and do not flower (8.3). After overwintering, they flower early in the spring. At that time, the staminate catkins become more pendulous and elongate, 2–3 in. (5–7.6 cm) long, and open (8.4) to shed abundant wind-dispersed pollen. At the same time, the much smaller, 0.2–0.4 in. (0.5–1 cm) long-ovoid pistillate catkins open, turn red, and become receptive to pollen (8.4). After fertilization, and over the summer, the pistillate catkins grow into broad-ovoid green "cones" 0.4–0.6 in. (1–1.5 cm) long (8.1, 8.5) and contain maturing, small fruits. These "cones" then dry out, turn brown, shed their narrowly winged single-seeded fruits (samaras) in late fall and winter, and may remain on the plant as empty "cones" for another year (8.6).

OCCURRENCE · FACW This species occurs throughout our area. It is common in fens and the minerotrophic periphery of bogs. It also occurs on wet mineral soils, where it may form extensive thickets.

8.1 Speckled alder: branch with green pistillate "cones." 8.2 Speckled alder: part of branch in winter showing stalked buds. 8.3 Speckled alder: closed pistillate (arrow 1) and staminate (arrow 2) catkins at end of first summer, prior to overwintering. 8.4 Speckled alder: pistillate (arrow 1) and staminate (arrow 2) catkins after overwintering, and at flowering stage in early spring. 8.5 Speckled alder: green pistillate "cones" in summer, containing maturing fruits. 8.6 Speckled alder: pistillate "cones" in their second spring, after shedding most of their seeds in winter. Photograph: Glen H. Mittelhauser.

9

Arrowwood (*Viburnum dentatum*) is a deciduous shrub in the family Adoxaceae. Fast-growing specimens form long, straight stems that were used by Native Americans for the shafts of arrows.

IDENTIFICATION The plant is 3–12 ft. (0.9–3.7 m) tall, but typically shorter than 8 ft. (2.4 m) in the peatlands of our area. Its opposite leaves are coarsely toothed (9.1). Many small white flowers are grouped into a slightly convex inflorescence (9.2). The fruits (drupes) are blue black (9.3).

OCCURRENCE · FAC Arrowwood occurs throughout our area at open and open-wooded sites and borders in fens and the minerotrophic periphery of bogs. It also is found along lake and river shorelines and relatively dry upland borders.

9.1 Arrowwood: stem with leaves. **9.2** Arrowwood: inflorescence.
9.3 Arrowwood: fruits (drupes).

10

Clammy azalea

Clammy azalea This tall deciduous shrub, *Rhododendron viscosum*, is a member of the heath family, Ericaceae. "Clammy" in its name comes from the sticky glands on its flowers. It is also called swamp azalea. All parts of the plant are considered to be poisonous.

IDENTIFICATION Clammy azalea can grow 16 ft. (about 5 m) tall but rarely exceeds 10 ft. (3 m) and commonly is 3–7 ft. (about 1–2 m) tall in our peatlands. It is much branched, and its young twigs are more or less pubescent (covered with short, soft hairs). It has ovoid alternate leaves, 0.7–2.4 in. (2–6 cm) long, with smooth edges and a short point at the tip (10.1). The white, rarely pink, flower petals are fused into a tubular base and spread apart at the terminus, in all about an inch (2–3 cm) long, and are covered by sticky, glandular hairs (10.1, 10.2). The style and stamens project well beyond the petals. The species flowers after the expansion of the leaves, June through July in New England. The fruit is an elongate capsule, 0.4–0.8 in. (1–2 cm) long, green at first, maturing to brown (10.3) before splitting open to release the seeds, and covered by stiff hairs. Another species, pink azalea (*R. periclymenoides*—FAC) differs from clammy azalea in its earlier flowering, before or during the expansion of the leaves, May through June in New England, and has pink (rarely white) flowers, and larger leaves.

OCCURRENCE · FACW Clammy azalea occurs as an understory shrub in swamps and wooded fens and also in shrubby wetlands in southeastern and eastern United States. In our area it is found only in Pennsylvania, New Jersey, New York, and the six New England states, becoming rare at its northeastern limit in southwestern Maine. Pink azalea's wetland habitats and distribution are similar, except that it more frequently occurs in upland forests and is more extensively distributed in central United States. It is found in all our states except Indiana, Michigan, and Wisconsin. Both species are absent from Canada.

10.3

10.1 Clammy azalea: top of branch in flower. Photograph: Arthur Haines.
10.2 Clammy azalea: flower and flower buds showing sticky hairs. Photograph:
Arthur Haines. **10.3** Clammy azalea: capsular fruits and bud. Photograph:
Joshua Fecteau.

11

Bog birch

Although most birch species in our region are deciduous trees, this species, *Betula pumila*, is a deciduous shrub. It is a member of the birch family, Betulaceae. Bog birch is sometimes called dwarf birch, but that name is also applied to two other species of shrub birches in our region.

IDENTIFICATION Bog birch grows to a height of 3–13 ft. (about 1–4 m) but typically grows to 3–6 ft. (roughly 1–2 m) tall in our peatlands (11.1). It is a bushy shrub with smooth dark bark, except for warty white glands on twigs that may vary in abundance from none to many (11.2 to 11.4). The alternate leaves have short petioles and broadly ovoid blades. These blades are 1–2 in. long (exceptionally to 2.7 in.) (2.5–5 cm, exceptionally to 7 cm) and are broadly acute to rounded at the apex (11.2, 11.3). These somewhat thick and leathery leaves have two to six pairs of lateral veins, edges that are coarsely toothed (the teeth may be rounded), and are often whitish underneath (11.2, 11.3). Pubescence on the leaves varies from none to dense. The separate pistillate and staminate catkins are 0.3–0.8 in. (about 1–2 cm) long (11.3, 11.4) and are produced on the same plant. The pistillate catkins produce small samaras with wings slightly narrower than the body. These fruits are released in fall and winter. Bog birch hybridizes with two tree-birch species in our region, paper and yellow birch, *B. papyrifera* and *B. alleghaniensis*, respectively. The hybrids have intermediate characteristics and may be expected where these tree species grow close to *B. pumila*.

OCCURRENCE · OBL Bog birch occurs in all our states and provinces except Pennsylvania and Rhode Island. Its eastern North American southern limit is in our southern tier of states from Illinois to New Jersey. In these states it is variously listed as extremely rare to rare, endangered, threatened, or of special concern. Northeastward in New York and the New England states it has the same listings, depending on the state. It grows in circumneutral and alkaline fens and the minerotrophic periphery of bogs, along lake and river shorelines, and often in association with northern white cedar (*Thuja occidentalis*). Bog birch is most common and widespread in the peatlands of the western parts of our area where calcareous geological deposits are more common.

11.1 Bog birch: stand of plants in a fen. Photograph: Russ Schipper. 11.2 Bog birch: leafy twig, with pistillate catkin on right. Photograph: Arthur Haines. 11.3 Bog birch: leafy twig with pistillate catkins. Photograph: Arthur Haines. 11.4 Bog birch: twig with staminate catkins at leaf-out time in the spring. Photograph: Warren H. Wagner Jr. Used by permission of the University of Michigan Herbarium.

12

Highbush blueberry

This species, *Vaccinium corymbosum*, is a tall deciduous shrub in the heath family, Ericaceae. Cultivars of the species are widely grown for their berries.

IDENTIFICATION This shrub grows to 13 ft. (4 m) tall, but in our area, the northern part of its range, it rarely is taller than 10 ft. (3 m) (12.1). The bark of its larger stems splits longitudinally and peels (12.2). The leaves are alternate. The white flowers, sometimes tinged pink, are urn shaped and occur in dense clusters (12.3). The berries ripen from green to white to blue (12.4). Maleberry (*Lyonia ligustrina*—FACW) (Haines 2011) may be confused with

highbush blueberry because of its similar vegetative and floral appearances and habitat, but it has dry fruits (capsules) instead of berries, and it does not occur in the most northern and northwestern parts of our region.

OCCURRENCE · FACW Highbush blueberry occurs widely in wetlands, including peatlands, throughout our area and well beyond it to the south, but becomes less abundant in the area's northernmost latitudes. It is most common at wooded and open-wooded fens and wet bog borders. It also occurs at acidic lake and stream shorelines and at rocky uplands.

12.1 Highbush blueberry: a plant in wooded fen understory. **12.2** Highbush blueberry: stem showing distinctive bark. **12.3** Highbush blueberry: flowers. **12.4** Highbush blueberry: fruits.

13

Alder-leaved buckthorn (*Rhamnus alnifolia*) is a native deciduous shrub in the family Rhamnaceae. It may easily be confused with the exotic invasive shrub European buckthorn (*R. cathartica*).

IDENTIFICATION *Rhamnus alnifolia* shrubs are commonly less than 40 in. (1.0 m) tall but can be as tall as 70 in. (1.8 m). Leaves are alternate and have arcuate veins (13.1, 13.2). Dogwood leaves also have arcuate veins, but those leaves are opposite. Unlike the exotic, invasive, and often taller *R. cathartica* (FAC), *R. alnifolia* has no spines. *Rhamnus alnifolia* flowers are only 0.16–0.25 in. (4–6 mm) wide and have greenish-yellow sepals. These flowers are either pistillate (13.2) or staminate and occur on separate plants. The 0.25–0.30 in. (6–8 mm) fruits (drupes) ripen from green to red to black (13.3). The flowers of *R. cathartica* have four sepals and four petals, whereas those of *R. alnifolia* have five sepals and no petals (13.2).

OCCURRENCE · OBL Alder-leaved buckthorn occurs in fens, minerotrophic peripheral areas of bogs, and lake and river shorelines in all our jurisdictions. European buckthorn (FAC) also occurs throughout our area in a wide range of habitats, most commonly in anthropogenic habitats like roadsides, fencerows, and pastures, but also in various forest types and a range of disturbed and pristine wetlands, including circumneutral and alkaline fens.

13.1 Alder-leaved buckthorn: leaves. **13.2** Alder-leaved buckthorn: pistillate flowers with very short, infertile stamens. Photograph: Warren H. Wagner Jr. Used by permission of the University of Michigan Herbarium. **13.3** Alder-leaved buckthorn: fruits (drupes). Photograph: Greg Vaclavek.

14

Red-osier dogwood (*Swida sericea*) is a deciduous shrub in the Cornaceae or dogwood family.

IDENTIFICATION This shrub can be as tall as 10 ft. (3.0 m) but typically is shorter than 8 ft. (2.4 m) in our peatlands. Its leaves are opposite on a red stem and have arcuate veins (14.1, 14.2). It is distinguished from another dogwood of our wetlands, silky dogwood (*S. amomum*—FACW) by its red stem and white pith. It is distinguished from alder-leaved buckthorn, which also has leaves with arcuate veins, by its opposite-leaf arrangement, red stem, and other features. Red-osier dogwood has numerous small four-petaled white flowers in a flat to slightly convex inflorescence (14.2) and has white fruits (drupes) (14.3).

OCCURRENCE · FACW This shrub occurs at open and sparsely wooded sites in our fens and minerotrophic peripheral areas of our bogs. It is widespread and common in these and other wetlands, along shores, and in wet ditches throughout our area.

14.1 Red-osier dogwood: stem with opposite leaves. **14.2** Red-osier dogwood: inflorescence. **14.3** Red-osier dogwood: fruits (drupes).

15

Mountain holly (*Ilex mucronata*, formerly *Nemopanthus mucronatus*) is a deciduous shrub in the holly family, Aquifoliaceae.

IDENTIFICATION This shrub may be as tall as 10 ft. (3.0 m) but typically is 3–7 ft. (0.9–2.1 m) in our peatlands, except only 1–3 ft. (0.3–2.1 m) at ombrotrophic sites. Young twigs and leaf petioles are purple (15.1). The leaves are alternate, smooth edged, and have fine needlelike (mucronate) tips (15.1, 15.2). The flowers are only 0.15–0.2 in. (0.4–0.5 cm) across, white and green, on long stalks, open in spring when the young leaves are still expanding (15.2), and emit a pleasant fragrance. Fruits (drupes) mature in late summer, becoming an intensely saturated, slightly purplish red color (15.1).

OCCURRENCE · OBL This species occurs throughout our area in wetlands and along shores. It grows at both ombrotrophic and minerotrophic areas of peatlands and at both open and wooded sites. Mountain holly also grows in montane spruce-fir forests, including subalpine forests on our mountains. Some of these sites may not qualify as wetlands, despite the official designation of OBL for this species.

15.1 Mountain holly: branch with leaves showing purple petioles, and fruits (drupes). **15.2** Mountain holly: branch in flower, with recently emerged and still-expanding leaves.

16

White meadowsweet (*Spiraea alba* var. *latifolia*) is a deciduous shrub in the rose family, Rosaceae.

IDENTIFICATION This shrub can be as tall as 6.5 ft. (2 m) but typically is shorter than 5 ft. (1.5 m). It has alternate, coarsely toothed, simple leaves. The small (less than or about 0.4 in. or 1.0 cm) white and pink flowers have numerous stamens and occur in typically conical, taller-than-wide, and showy inflorescences (16.1, 16.2). The fruits (follicles) stay green into the fall (16.3) but over the winter dry out, turn brown, and split open to release seeds.

OCCURRENCE · FACW This species occurs throughout our area, except Wisconsin and Indiana, in fens and the peripheral, minerotrophic parts of bogs. It prefers unwooded or sparsely wooded sites. It also occurs widely at non-wetland fields, roadsides, and other habitats.

16.1 White meadowsweet: top of shrub with inflorescences. **16.2** White meadowsweet: close-up of an inflorescence. **16.3** White meadowsweet: fruits (follicles) in late October in central Maine.

17   **Poison-sumac** This deciduous shrub or small tree, *Toxicodendron vernix*, formerly *Rhus vernix*, is a member of the sumac or cashew family, Anacardiaceae. This family contains poisonous species of tropical and temperate regions, including poison-ivy, *T. radicans*. All parts of the poison-sumac plant are very poisonous by touch to humans, but the fruits are fed upon by many species of birds and mammals, and all parts of the plant by insects.

IDENTIFICATION Although I include this species with tall shrubs, rather than trees, it may grow to about 30 ft. (approx. 9 m) tall under most favorable conditions. However, typically it is no taller than 16 ft. (about 5 m) in our peatlands. It is often branched from the base but may have a single trunk, and has gray, near-smooth bark. Each of its alternate, compound leaves has a reddish rachis bearing 7 to 13 long-ovoid and pointed leaflets with smooth edges (17.1). The leaves turn a brilliant red in autumn (17.2). Inflorescences as long as 8 in. (20 cm) emerge from the leaf axils, are spreading, and may hang down, and have small white, yellow, or green flowers (17.1, 17.3). The globular fruits (drupes) are gray to white and 0.15–0.2 in. (about 4–5 mm) in diameter (17.4). The lower-growing or vine-like poison-ivy (*Toxicodendron radicans*—FAC) also grows in our peatlands, has only three leaflets per leaf, and is nearly impossible to confuse with poison-sumac. Other sumac shrubs, staghorn sumac (*Rhus hirta*) and smooth sumac (*R. glabra*), have bright red fruits and rarely grow in wetlands.

OCCURRENCE · OBL Poison-sumac occurs in all our states and provinces except New Brunswick and Prince Edward Island. It is a wetland species that grows in shrubby borders of circumneutral and alkaline fens, fen thickets, open-wooded fens, and the minerotrophic edges of bogs, especially in calcareous terrain. It also occurs in mineral-soil marshes, swamps, and shrub wetlands. Poison-ivy occurs in all our states and provinces in wetland habitats similar to those of poison-sumac but also ranges widely in non-wetland habitats.

17.1 Poison sumac: top of plant in bloom. Photograph: Robert W. Smith. **17.2** Poison sumac: plants in the fall. Photograph: Anton A. Reznicek. **17.3** Poison sumac: part of inflorescence with flowers. Photograph: Robert W. Smith. **17.4** Poison sumac: fruits (drupes) in winter. Photograph: Donald S. Cameron.

18

Rhodora This species, *Rhododendron canadense,* is a deciduous shrub in the heath family, Ericaceae. Commonly it is the most showy early blooming shrub of our peatlands.

IDENTIFICATION Rhodora grows to about 50 in. (about 1.3 m) tall but typically is 20–40 in. (0.5–1.0 m) tall in our peatlands. The gray-green leaves are alternate (18.1). Flower and leaf buds open in May in much of our region. The complex flowers are fully formed when the leaves are still quite small. These flowers are at the ends of the stems, are rose purple with white at the base of some of the petals (18.2, 18.3), and are all white in one form of the species. The fruits (capsules) are finely hairy and pinkish when young (18.1), but in the fall they dry out, turn gray and brown, and split open to shed their seeds (18.4).

OCCURRENCE · FACW Rhodora is absent from Wisconsin, Illinois, Michigan, Indiana, and Ohio but occurs in all the other states and provinces of our region. It grows at open and sparsely wooded sites and forest edges in bogs and fens, at peaty lake and stream shorelines, and on mineral soil in meadows, wet pastures, and rocky slopes.

18.1

18.1 Rhodora: shrub in leaf with young, hairy fruits (capsules). **18.2** Rhodora: plant in full flower as leaves are starting to emerge in late May in central Maine. **18.3** Rhodora: close-up of flowers. **18.4** Rhodora: fruits (capsules) starting to split open at the end of October in central Maine.

19

Sweetgale (*Myrica gale*), in the Myricaceae family, is an aromatic deciduous shrub. It has been used for insect repellent, perfume, flavoring drinks, seasoning, and medicine. It has nitrogen-fixing bacteria in root nodules and can increase the nitrogen content of the soil where it grows.

IDENTIFICATION Although sweetgale can be 7 ft. (2.1 m) tall when fully grown, most individuals in the peatlands of our area rarely become taller than 5 ft. (1.5 m). The gray-green leaves are alternate, widest near the tip, and gradually narrow to where they join the stem, virtually without a petiole. The distal third of the leaf is toothed (19.1). The plant flowers in the spring before the leaves emerge. The staminate and pistillate catkins are on separate plants. The staminate catkins are more or less cylindric, 0.4–0.8 in. (1–2 cm) long, and the flowers are separated by prominent scales (19.2). The pistillate catkins are ovoid, about 0.4 in. (approx. 1 cm) long, with flowers separated by scales, each flower with a bifid stigma (19.3). At the fruiting stage, a pistillate catkin forms a compact group of nutlets (19.4). When bruised, the plant emits a spicy-sweet fragrance.

19.1

OCCURRENCE · OBL Sweetgale occurs throughout our area except in Illinois, Indiana, and Ohio. It grows in minerotrophic and unwooded or sparsely wooded parts of peatlands and forms extensive stands in wet parts of some open fens, often where the fens border open water. The species also commonly grows on mineral soils along lake and stream shorelines.

19.1 Sweetgale: part of shrub with leaves and pistillate catkins in fruit.
19.2 Sweetgale: stem prior to leaf emergence, with several staminate catkins showing anthers. Photograph: Marilee Lovit. **19.3** Sweetgale: stem prior to leaf emergence, with pistillate catkins. Red bifid stigmas emerge from the flowers. Photograph: Marilee Lovit. **19.4** Sweetgale: stem with pistillate catkins with fruits (nutlets) at mid-stage of development.

20

Bog willow This species, *Salix pedicellaris*, is a deciduous shrub in the willow family, Salicaceae. A few other shrub willow species occur sporadically in our peatlands, but this species is the most common and regularly occurring one.

IDENTIFICATION Bog willow is a distinctive willow in its complete lack of pubescence. It has smooth leaf margins, and long-stalked catkins. The loosely branching shrub may grow about 60 in. (1.5 m) high but commonly reaches only about 40 in. (1.0 m) in our peatlands. The elongate-ovoid, alternate leaves have blades 0.8–2.4 in. (2–6 cm) long. When young, these blades have a whitish waxy coating on the underside (20.1). As with all willows, the winter buds are capped by only a single scale. The smaller branches are flexible and not easily broken. The flowers occur in catkins (aments) about 0.4–1.2 in. (1–3 cm) long, with staminate (20.2) and pistillate (20.3) catkins on separate plants. The fruit is a capsule that splits open to release its fluffy seeds to the wind (20.3).

OCCURRENCE · OBL Bog willow occurs in all 19 of our states and provinces. It is a northern species whose distribution extends into Arctic Canada. It is generally rare near the southern fringes of our area, where it reaches its southern limit. It is a wetland species characteristic of circumneutral and alkaline fens and the minerotrophic periphery of bogs in calcium-rich terrain. As these conditions occur more widely at the western half of our area, the species is more abundant in peatlands there than in the east. It also occurs at lake and river shorelines and swamps.

20.1 Bog willow: twig with young leaves showing whitish waxy bloom on undersides of leaves. Photograph: Donald S. Cameron. **20.2** Bog willow: leafy twig with staminate catkin (ament). Photograph: Robert Routledge. **20.3** Bog willow: leafy branch with pistillate catkins (aments). Some of the capsules are starting to open, revealing the white, fluffy seeds. Photograph: Anton A. Reznicek.

21

Common winterberry or black alder (*Ilex verticillata*) is a deciduous shrub in the holly family, Aquifoliaceae. The bright fruits are retained after the leaves fall, and the leafless fruit-laden branches are used for winter decoration.

IDENTIFICATION The shrub may be as tall as 13 ft. (4 m) but commonly is less than 8 ft. (2.4 m) tall in our peatlands. The rather stiff, alternate leaves are finely and sharply serrate (21.1 to 21.3) and dull on the underside. The small, white staminate (21.2) and pistillate (21.3) flowers are borne on short (0.04–0.16 in.; 1–4 mm) stalks, and on separate plants. The fruits (drupes) mature from green to orange to deep red but may retain an orange tinge (21.4). Smooth winterberry (*I. laevigata*—OBL) occurs in similar wetland habitats and can be difficult to distinguish from common winterberry. Smooth winterberry has staminate flowers on longer stalks, and leaves with a shiny underside.

OCCURRENCE · FACW Common winterberry grows in peatlands and mineral-soil wetlands and along lake, river, and stream shorelines throughout our area. It occurs in shrubby fens, the understory of wooded fens, and the peripheral, minerotrophic parts of bogs.

21.4

21.1 Common winterberry: shrub in fruit prior to leaf-fall at beginning of October in central Maine. **21.2** Common winterberry: stem with staminate flowers. **21.3** Common winterberry: stem with pistillate flowers. **21.4** Common winterberry: stem with fruits (drupes) after leaf-fall in late October in central Maine.

22

Withe rod or wild raisin (*Viburnum nudum* var. *cassinoides*) is a deciduous shrub with edible fruit in the family Adoxaceae.

IDENTIFICATION The plant grows as tall as 13 ft. (4 m) but is generally shorter than 8 ft. (2.4 m) in our area and is stunted to 2–4 ft. (0.5–1.2 m) at ombrotrophic areas of bogs. The leaves are leathery and opposite on scurfy twigs (22.1). The buds are naked and reddish tan. The terminal bud is 0.4–0.7 in. (1.0–1.8 cm) long (22.2) and produces the terminal inflorescence. Many small white or off-white flowers occur in the slightly convex inflorescence (22.3). The fruits (drupes) ripen from white to pink to red to blue black (22.1).

OCCURRENCE · FACW Withe rod occurs in all the states and provinces of our area in fens, bogs, and mineral-soil wetlands at open and sparsely wooded sites and edges. It is also widespread at non-wetland thickets, clearings, and forest edges.

22.1 Withe rod: stem with opposite leaves, and fruits (drupes). **22.2** Withe rod: tip of stem with long terminal bud flanked by two axillary buds. **22.3** Withe rod: stem with inflorescence.

23

Dwarf red blackberry (*Rubus pubescens*), also called dwarf raspberry, is a member of the rose family.

IDENTIFICATION Although rooted basal stems are woody, the abundant trailing upper stems are unusual for *Rubus* in being herbaceous and lacking prickles. The stems form runners and may have fine hairs in varying amounts. The alternate, compound leaves have 3 leaflets, with center leaflet typically diamond-shaped (22.1). The flower's five white petals are borne on short stalks (23.2). The so-called berry is an aggregate of drupelets, light green to white when young but dark red and tasty at maturity (23.1). When blackberries are picked, the receptacle which bears the drupelets stays with the berry, whereas in raspberries the receptacle stays behind on the stem. In this species, the berries come off the receptacle only with difficulty.

OCCURRENCE · FACW Found in fens, minerotrophic edges of bogs, and a range of other moist sites, typically in partial shade. Occurs throughout our area.

23.2

23.1 Dwarf red blackberry: a trailing plant with "berries" (aggregates of drupelets).
23.2 Dwarf red blackberry: flower.

.

24

Velvet-leaved blueberry (*Vaccinium myrtilloides*) is a deciduous shrub in the heath family, Ericaceae.

IDENTIFICATION The plant is 5–35 in. (13–89 cm) high, typically 8–18 in. (20–46 cm) in our area. The leaves are smooth-edged, alternate, and bear fine, short hairs on the underside, edges, and along veins on the upper side (24.1 to 24.3). The leaves are commonly blue green or gray green like sage (24.3) but can be yellow- or red-tinged (24.1, 24.2). The flowers are globose urn-shaped, green to white, or white tinged with purple (24.1, 24.2). The fruit is a dark blue berry with a white bloom (24.3) and is sour compared to the common low-bush blueberry of commerce, *V. angustifolium* (FACU). The latter species also occurs naturally in peatlands and is distinguished from *V. myrtilloides* by leaves that have a finely serrate edge and a lack of pubescence, except rarely a sparse pubescence along the veins.

OCCURRENCE · FACW *Vaccinium myrtilloides* is occasional on hummocks at open and wooded sites in bogs and fens. Also it occurs at non-wetlands at a wide range of open and wooded habitats. It becomes rare in the southern parts of our area and is absent from New Jersey and Rhode Island. *Vaccinium angustifolium* occurs in similar habitats in all our states and provinces but is more widely abundant in non-wetland habitats.

24.1 Velvet-leaved blueberry: leafy stem with flowers. **24.2** Velvet-leaved blueberry: flowers. **24.3** Velvet-leaved blueberry: fruits (berries).

25

Black chokeberry (*Aronia melanocarpa*) is a deciduous shrub in the rose family (Rosaceae) with fruits that are bitter or astringent.

IDENTIFICATION Although this species can grow to 10 ft. (3 m) tall, I include it among the short and dwarf shrubs because in most of our peatlands it is shorter than 40 in. (1 m). The leaves are alternate and have fine, rounded, and/or pointed teeth along the edge (25.1, 25.2). The leaf undersides and twigs have few if any hairs on them. The flowers are 0.4–0.5 in. (1.0–1.4 cm) wide, have five white petals, sometimes tinged pink, and each petal is borne on a short stalk (25.1). Like its relative, apple, the fruits are pomes (fleshy part not derived from ovary). These pomes mature to a black or dark purple color (25.2). The similar purple chokeberry (*A. floribunda, A. prunifolia*, and other Latin names—FACW), which also occurs in peatlands, has pomes that are lighter purple at maturity, branchlets and leaf undersides that are sparsely to abundantly pubescent, and commonly the plant is taller.

OCCURRENCE · FAC Black chokeberry is common in open, shrubby, and sparsely wooded parts of bogs and fens throughout our area. It also occurs in non-wetland thickets and open woods, including at rocky balds. Purple chokeberry also occurs throughout our area, more in wetlands than non-wetlands compared to black chokeberry.

25.2

25.1 Black chokeberry: branch in flower.
25.2 Black chokeberry: fruits (pomes).

26

Mountain honeysuckle or mountain fly honeysuckle (*Lonicera villosa*) is a deciduous shrub in the honeysuckle family, Caprifoliaceae.

IDENTIFICATION The plant is 15–40 in. (0.4–1.0 m) tall, has opposite gray-green leaves (26.1 to 26.3), and buds with two scales (26.1, 26.2). Pubescent (26.1 to 26.3) and non-pubescent varieties occur in our area. The flowers are white (26.3) to white tinged with yellow, or entirely yellow, and occur in pairs appearing to share a single ovary, although what appears single is really two ovaries in a single fleshy cup not derived from the ovary. This structure is subtended by bractlets (26.3). The fruit resembles a berry but is an accessory fruit in which the fleshy part is derived the above-mentioned fleshy cup, and is therefore a faux berry. It is 0.3–0.4 in. (0.7–1.0 cm) long, blue when ripe (26.4), and widely considered to be edible, although a mild toxicity has been reported.

OCCURRENCE · FACW Except for Illinois, Indiana, and New Jersey, the species occurs throughout our area. It grows at wooded and open areas of fens and at minerotrophic, peripheral areas of bogs. It also occurs on moist-wet mineral soils in meadows and forests and at open subalpine and alpine sites. It is rare at the southern fringes of its range in our area.

26.2

26.4

26.1 Mountain honeysuckle: branches of shrub in leaf. **26.2** Mountain honeysuckle: detail of end of stem including buds. **26.3** Mountain honeysuckle: stem with a pair of flowers. **26.4** Mountain honeysuckle: faux fruit close-up, showing the two places where flower parts had been attached. Photograph: Marilee Lovit.

27

Black huckleberry This species, *Gaylussacia baccata*, is a deciduous shrub in the heath family, Ericaceae. The black fruits are sweet and popular for pies and other culinary uses. In the fall when the leaves turn, the plants color parts of our peatlands a vivid red.

IDENTIFICATION The plant typically is 20–40 in. (0.5–1.0 m) tall. The alternate leaves are resin dotted on both sides and sticky when squeezed between the fingers (27.1). The leaves turn bright red in the fall (27.2). The urn-shaped flowers are more elongate than those of highbush blueberry, are dull red or white with a blush of rusty red, and are covered by resin specks (27.3). The fruits (berry-like drupes) mature from green to black (27.4) and have seeds that are harder than those of blueberry. A less widespread species in our peatlands, dwarf huckleberry (*G. bigeloviana*—OBL), is only 8–20 in. (0.2–0.5 m) tall, mostly is covered by tiny hairlike glands, and has leaves that are more lustrous on the underside.

OCCURRENCE · FACU Black huckleberry occurs throughout our area in open and open-wooded parts of fens and bogs and is abundant at some relatively dry uplands. Dwarf huckleberry is restricted to peatlands and heathlands and occurs in our states and provinces except Vermont, Ohio, Indiana, Illinois, Michigan, and Wisconsin.

27.2

27.1 Black huckleberry: branch with leaves and flowers. **27.2** Black huckleberry: leaves in fall color, backlit. **27.3** Black huckleberry: close-up of inflorescence. **27.4** Black huckleberry: fruits (berrylike drupes).

28

Labrador tea This species, *Rhododendron groenlandicum*, is an evergreen shrub in the heath family, Ericaceae. Its aromatic leaves have been used to make tea and medicine, but toxicity has been reported for concentrated doses (http://plants.usda.gov /plantguide/pdf/cs_legr.pdf).

IDENTIFICATION The plant is 12–40 in. (0.3–1.0 m) tall. The twigs are densely hairy, and the leaves are alternate and have revolute margins (28.1 to 28.3). The underside of new leaves is white to-mentose (woolly), and the upper side has white resinous dots. With age, the wool turns rusty colored, and the resinous dots disappear. The plant has white flowers in inflorescences at the end of stems (28.1, 28.2). As fruits form, the stem grows beyond the original terminal position of the inflorescence, as indicated in photo 28.3 by the position of the current year's fruits. The fruits (capsules) are slender, dry out, and split open to shed their seeds.

OCCURRENCE · OBL Labrador tea occurs throughout our area except Illinois and Indiana. It is sporadic at the southern fringes of the region. Farther north, it is common in bogs and acidic fens at both open and wooded sites. It also occurs along peaty lake and stream shorelines and in peaty pockets on open slopes and ridges, often rocky, in the mountains.

28.1 Labrador tea: stems, one with an inflorescence. **28.2** Labrador tea: stem and inflorescence detail. **28.3** Labrador tea: branch with fruits (capsules).

29

Bog laurel (*Kalmia polifolia*), or bog American laurel, is an evergreen shrub in the heath family, Ericaceae.

IDENTIFICATION When not in flower, this short and slender plant is inconspicuous in its usual setting, shrub heath vegetation, even when common. It typically is only 6–14 in. (15–36 cm) tall, but may reach 2 ft. (about 0.6 m) tall under the most favorable conditions. The opposite leaves are typically narrow (linear) (29.1, 29.2) but vary in width to ovate (29.3) and are whitened underneath. The stem has a pair of longitudinal ridges between each pair of leaves, so that a cross-section of the stem would appear round with two bumps directly across from each other. The successive pairs of leaves and ridges shift by 90 degrees, so when one looks directly down on the top of the plant, the leaves form a plus (+). The rose-purple or pink flowers cluster at the end of the stem (29.1, 29.2) and give rise to fruits (capsules) that split into their five chambers (locules) and turn brown and dry at maturity (29.3).

OCCURRENCE · OBL Bog laurel occurs throughout our area except Illinois, Indiana, and Ohio. It grows in open areas of bogs and acidic fens and on peaty soil high on mountains.

29.1 Bog laurel: branch in flower. A typically slender plant appears on the right.
29.2 Bog laurel: close-up of an inflorescence. **29.3** Bog laurel: plant in fruit (capsules, each showing five locules). The leaves on the fruiting plant are much narrower than a pair of leaves on the bog laurel plant just below it. The leaves at upper right belong to a leatherleaf plant.

30

Sheep laurel, sheep American laurel, or lambkill (*Kalmia angustifolia*) is an evergreen shrub in the heath family (Ericaceae) whose leaves are poisonous to livestock.

IDENTIFICATION The plant is 8–40 in. (0.2–1.0 m) tall in our area, and at the short end of that range at ombrotrophic areas of bogs. The twigs lack ridges, are circular in cross section, and bear naked buds. The leaves typically are whorled three to a node, or occasionally opposite, with older leaves bending down on curved stalks (30.1, 30.2). Inflorescences form from buds in the axils of the prior year's leaves and appear lateral as the new year's stems grow out. The flowers are reddish purple to deep pink (30.1), rarely white. Fruits (capsules) are only 0.1–0.2 in. (0.25–0.5 cm) wide and on curved stalks (30.2). Fruits darken from green to tan (30.2) to brown (and dry) at maturity.

OCCURRENCE · FAC Sheep laurel occurs throughout our area except Wisconsin, Illinois, Indiana, and Ohio. It grows in a wide variety of wetland and non-wetland habitats at both open and wooded sites. It is most abundant on acidic soils, including sterile gravel and the peats of our bogs and acidic fens.

30.2

30.1 Sheep laurel: stems in flower.

30.2 Sheep laurel: stems with fruits (capsules).

31

Leatherleaf (*Chamaedaphne calyculata*) is an evergreen shrub in the heath family, Ericaceae. At open areas of bogs and acidic fens in our area it is commonly the most abundant species of low shrub.

IDENTIFICATION In our peatlands, leatherleaf can grow to 55 in. (1.4 m) tall under ideal minerotrophic conditions, but generally is 15–30 in. (0.4–0.75 m) tall. At wet, ombrotrophic bog sites, it may be severely stunted and not exceed 10 in. (0.25 m). The leaves are alternate, often point upward, are thick and leathery, and decrease in size toward the tip of the stem (31.1 to 31.3). The leaves have minute scales that appear as dots to the naked eye. These scales commonly are white on both sides of the leaf (31.1, 31.3) but frequently are rusty brown on the underside (31.2). The live, evergreen leaves may become brownish in winter. The urn-shaped white flowers are borne individually in the axils of small leaves and hang down (31.1). The fruits (capsules) are green when young (31.3) but turn brown and dry at maturity.

31.3

OCCURRENCE · OBL Leatherleaf occurs at open and wooded parts of peatlands throughout our area and at mineral-soil wetlands, lake and stream shorelines, and in floodplains.

31.1 Leatherleaf: stem with flowers beginning to wither. **31.2** Leatherleaf: stem with buds, showing rusty-brown scales on leaf undersides. **31.3** Leatherleaf: horizontal stem with young fruits (capsules), as seen from above.

32

Rosy meadowsweet, steeplebush, or hardhack (*Spiraea tomentosa*) is a shrub in the rose family, Rosaceae. It attracts many butterflies and moths.

IDENTIFICATION This species has a single or sparsely branched stem that can be as tall as 40 in. (1 m) but typically is 15–30 in. (about 0.4–0.8 m) tall (32.1). The alternate, 1.2-to-2-in. (3–5 cm) long ovoid to lanceolate leaves have toothed edges, short petioles, firm texture, undersides that are white, yellowish, or reddish-brown tomentose (woolly), and are prominently veined (32.2). The terminal inflorescence is steeple shaped (32.1), 2–8 in. (about 5–20 cm) tall, and densely packed with roseate flowers (32.3). The fruit is a follicle (32.4).

OCCURRENCE · FACW Rosy meadowsweet occurs in all 19 of our states and provinces. It grows in fens and the minerotrophic edges of bogs, more commonly in our central and western peatlands than in the east. Additionally, it grows in pastures, fields, meadows, at edges of mineral-soil wetlands and semi-open swamps, and along the shores of rivers and lakes.

32.1 Rosy meadowsweet: plants by lake edge. Photograph: Donald S. Cameron. **32.2** Rosy meadowsweet: leaves with tomentum (white, in this specimen) and prominent veins on underside. Photograph: Leanne Wallis. **32.3** Rosy meadowsweet: close-up of inflorescence. Photograph: Robert W. Smith. **32.4** Rosy meadowsweet: follicle fruits. Photograph: Russ Schipper.

33

Shining rose (*Rosa nitida*), also called bristly rose, is a deciduous shrub in the rose family, Rosaceae.

IDENTIFICATION The plants typically are 20–30 in. (0.5–0.75 m) tall, to a maximum of 40 in. (1 m). The stems (33.1) have fine, near-straight prickles, but no especially broad and curved ones at the nodes as in the similar, somewhat taller wetland species, swamp rose (*R. palustris*—OBL). The leaves are compound, commonly with nine (some with seven or five) serrate leaflets per leaf (33.1). The underside of leaflets is lustrous. It is dull or only slightly lustrous in *R. palustris*. In *R. nitida*, rose-colored flowers occur individually (33.2), but in *R. palustris* two or more occur together. The dry fruits (achenes) are enclosed in a fleshy or pulpy, subglobose red "hip" (33.3). The hip is not derived from the pistil, so the entire structure with the true fruits inside may be considered an accessory fruit.

OCCURRENCE · FACW Shining rose is a northeastern species that is absent from Pennsylvania, Indiana, Illinois, Michigan, and Wisconsin, and is rare in the southernmost parts of its range elsewhere in our area. In peatlands, it grows in the shrubby understory at wooded and open-wooded sites in fens and at the minerotrophic periphery of bogs, and elsewhere in wet thickets and along lake and stream shorelines. Swamp rose occurs in all our states and provinces in fens, the minerotrophic periphery of bogs, marshes, meadows and fields, on the shores of rivers, streams, and lakes, and in swamps and wet thickets.

33.1 Shining rose: prickly stems and compound leaves in fall. **33.2** Shining rose: flower. **33.3** Shining rose: the accessory fruit, called a hip (see text).

34

Bog rosemary (*Andromeda polifolia* var. *glaucophylla*) is an evergreen shrub in the heath family, Ericaceae. It resembles the culinary "herb" rosemary but is not similarly usable.

IDENTIFICATION The plant can reach 24 in. (61 cm) in height but generally is shorter than 18 in. (46 cm). The leaves are alternate, gray green (with a hint of blue), with netlike venation on top, and are white on the bottom (34.1, 34.2). The leaf edges are revolute (34.2). White to pink nodding flowers occur in small inflorescences at the ends of stems, each flower vase shaped (34.2). The young capsules (34.3) turn brown as they mature and dry.

OCCURRENCE · OBL This species frequently is abundant in open fens and bogs throughout our area. Typically it occurs at areas of wet peat moss, including at the edge of pools. Often it is most abundant at the outer edge of the open ombrotrophic area at raised bogs in the northern part of our area.

34.1

34.1 Bog rosemary: a stand of plants. **34.2** Bog rosemary: flowers. **34.3** Bog rosemary: capsules prior to browning and drying. Shriveled flowers (brown) that failed to form fruits are also shown.

35

Shrubby-cinquefoil
This species, *Dasiphora floribunda* or *D. fruticosa*, formerly in the genus *Potentilla*, is a deciduous shrub in the rose family, Rosaceae. It is commonly used as a decorative shrub in gardens, and there are several horticultural varieties.

IDENTIFICATION Shrubby-cinquefoil grows to a height of 1–3.3 ft. (0.3–1 m) and has shreddy bark on the larger stems. The alternate, compound leaves have five (less commonly seven) narrow leaflets, each leaflet 0.4–1.2 in. (1–3 cm) long, and more or less pubescent on both sides (35.1, 35.2). The winter bud is covered by two scales that meet along their edges. The bright yellow flowers have five petals, are solitary or bunched near the tip of a branch, and are 0.6–1.2 in. (1.5–3 cm) across (35.1, 35.3). Each flower produces multiple hairy achenes.

OCCURRENCE · FACW This species occurs in all our states and provinces except Rhode Island. It reaches its eastern North American southern limit in our southern tier of states from Illinois to New Jersey, and is listed as endangered in Pennsylvania. It grows in circumneutral and alkaline fens and the minerotrophic periphery of bogs, generally in the open, but also in thickets and at open-wooded sites. It is considered an indicator of alkaline conditions and is more abundant in peatlands at the western parts of our area where such conditions are more common. This species also occurs along lake, river, and stream shorelines, often rocky, and even in moist cracks of rock ledges and cliffs.

35.1 Shrubby-cinquefoil: plant in flower viewed from above. Photograph: Donald S. Cameron. **35.2** Shrubby-cinquefoil: close-up of leaves. Photograph: Arthur Haines. **35.3** Shrubby-cinquefoil: close-up of flowers. Photograph: Robert W. Smith.

35.1

35.2

35.3

36

Large cranberry (*Vaccinium macrocarpon*) is a native evergreen shrub in the heath family (Ericaceae) that is extensively cultivated for its berries.

IDENTIFICATION In the wild, the slender creeping stems bear 4-to-7-in.-tall (10–18 cm) ascending branches that bear flowers and fruits (36.1 to 36.3). The alternate elliptic-oblong leaves are 0.25–0.6 in. (0.6–1.5 cm) long, leathery, lustrous atop, and duller, paler or whitened beneath, and have rounded tips (36.1 to 36.3). The rosy-white flowers have turned-back petals (36.1, 36.2). In the wild, the red, acidic berries (36.3) are 0.4–0.6 in. (1.0–1.5 cm) in diameter. This species is distinguished from the small cranberry (*V. oxycoccus*) by its larger leaves with rounded tips, and its larger fruits. The generally different habitats give a hint on the species to expect but are not 100 percent reliable for making the distinction.

OCCURRENCE · OBL Large cranberry occurs throughout our area in peatlands at open, wet, minerotrophic sites, often in sphagnous fens where there is water flow. At ombrotrophic areas of bogs, it is largely restricted to pool edges. It also occurs on wet mineral soils, including at lakeshores.

36.1 Large cranberry: upright stems in flower. **36.2** Large cranberry: close-up of flowers. **36.3** Large cranberry: fruits (berries).

37

Small cranberry (*Vaccinium oxycoccos*) is an evergreen shrub in the heath family, Ericaceae.

IDENTIFICATION This dwarf creeping shrub has extremely slender branches that rarely ascend more than 3 in. (about 8 cm) above the surface (37.1). The alternate, ovate to long-triangular leaves are only 0.1–0.35 in. (2.5–9 mm) long, leathery, lustrous green atop, dull whitened below, with revolute edges and acute tips (37.1, 37.2). The rosy-white flowers have turned-back petals (37.1, 37.2). The mature sour fruits are mostly red, typically with one side pale or green (37.3), and 0.2–0.35 in. (0.5–0.9 cm) in diameter. This species is distinguished from the large cranberry (*V. macrocarpon*) by its more strictly prostrate habit, smaller and revolute leaves with pointed tips, and smaller fruits.

OCCURRENCE · OBL Small cranberry occurs throughout our area at open and open-wooded sites in bogs and acidic fens. It is common on peat moss at ombrotrophic areas of bogs but may not be noticed because of its small size and obscuration by taller plants. It also occurs on peaty non-wetland soils and along lakeshores.

37.1 Small cranberry: plants in flower, creeping over peat moss. **37.2** Small cranberry: close-up of a flower on a leafy stem. **37.3** Small cranberry: plants with fruits (berries), on red peat moss (*Sphagnum rubellum*).

38

Black crowberry (*Empetrum nigrum*) is a low creeping evergreen shrub in the heath family, Ericaceae.

IDENTIFICATION This shrub forms mat-like patches on surfaces where it spreads. The creeping plant may have ascending branches 2–4 in. (5–10 cm) tall and rarely to 6 in. (15 cm) tall. The alternate leaves resemble very short (0.1–0.2 in.; 2.5–5 mm) conifer needles (38.1). The stems may be minutely glandular, but not hairy. The tiny, purple to red to pink flowers are located at the leaf axils and may be perfect or imperfect, staminate (38.2), or pistillate (38.3) in various combinations on the same plant or separate plants, and these combinations may differ from site to site. The globose mature fruits (drupes) are 0.15–0.2 in. (0.4–0.6 cm) across and are dark blue to purple to black (38.4). Black crowberry is distinguished from red (purple) crowberry (*E. atropurpureum*—FAC) by its darker mature fruits and a lack of hairiness on its stems.

OCCURRENCE · FAC Black crowberry is a circumboreal and circumarctic species whose range in North America extends southward to include the following temperate areas and habitats. It occurs in Maine's coastal and subcoastal fens and bogs, largely east of Penobscot Bay and extending into coastal New Brunswick and Nova Scotia. In this same stretch along highly exposed rocky coastlines, it roots in peaty pockets and spreads over ledges. Also it is found in small subalpine and alpine fens in Maine, New Hampshire, Vermont, the Adirondack Mountains of New York, and the Shickshock (Chic-Choc) Mountains of Gaspé, Québec. At high elevations on these mountains it also grows in rocky terrain where it roots in peaty pockets and spreads over ledges. It is present in lowland peatlands and along rocky shores of the Upper Peninsula and Isle Royale of Michigan, and at the more northern latitudes of the parts of Ontario and Québec covered by this book. It is absent from our southern tier of states, Massachusetts, Connecticut, Rhode Island, New Jersey, Pennsylvania, Ohio, Indiana, and Illinois, and is also absent from Wisconsin.

38.1 Black crowberry: small and diffuse stand of plants. **38.2** Black crowberry: close-up of plant with imperfect, staminate flowers. **38.3** Black crowberry: close-up of plant with imperfect, pistillate flowers showing multiple-lobed stigmas. Photograph: Marilee Lovit. **38.4** Black crowberry: plants with fruits (drupes). Photograph: Glen H. Mittelhauser.

39

Creeping snowberry or creeping spicy-wintergreen

(*Gaultheria hispidula*) is a creeping/trailing evergreen shrub in the heath family, Ericaceae.

IDENTIFICATION This species creeps along the ground (39.1) and forms carpets under favorable conditions. The leaves are small (0.15–0.4 in.; 0.4–1.0 cm), shiny, and leathery, and alternate on the bristly, yellow-brown stem. Each small (about 0.15 in. or 0.4 cm) white flower is borne in a separate leaf axil (39.2). Flowers develop into small (about 0.2 in. or 0.5 cm), bristly, white edible "fruits" (39.1, 39.3) with wintergreen flavor. The true fruit is a thin-walled capsule, but it is surrounded by fleshy tissue derived from the sepals. When lacking these accessory fruits, the plant might be confused with one of the two cranberry species because of its similar-size leaves and creeping habit.

OCCURRENCE · FACW Except for Illinois and Indiana, this species occurs in all states and provinces of our area. In peatlands, mostly it is found on the shaded forest floor of evergreen wooded fen and bog, often in association with mosses. Additionally, it occurs widely in non-wetland moist mixed and conifer forests.

39.1 Creeping snowberry: plants in fruit. **39.2** Creeping snowberry: a stem in flower, held between fingers. Photograph: Glen H. Mittelhauser. **39.3** Creeping snowberry: close-up of stems with bristly accessory fruits (see text).

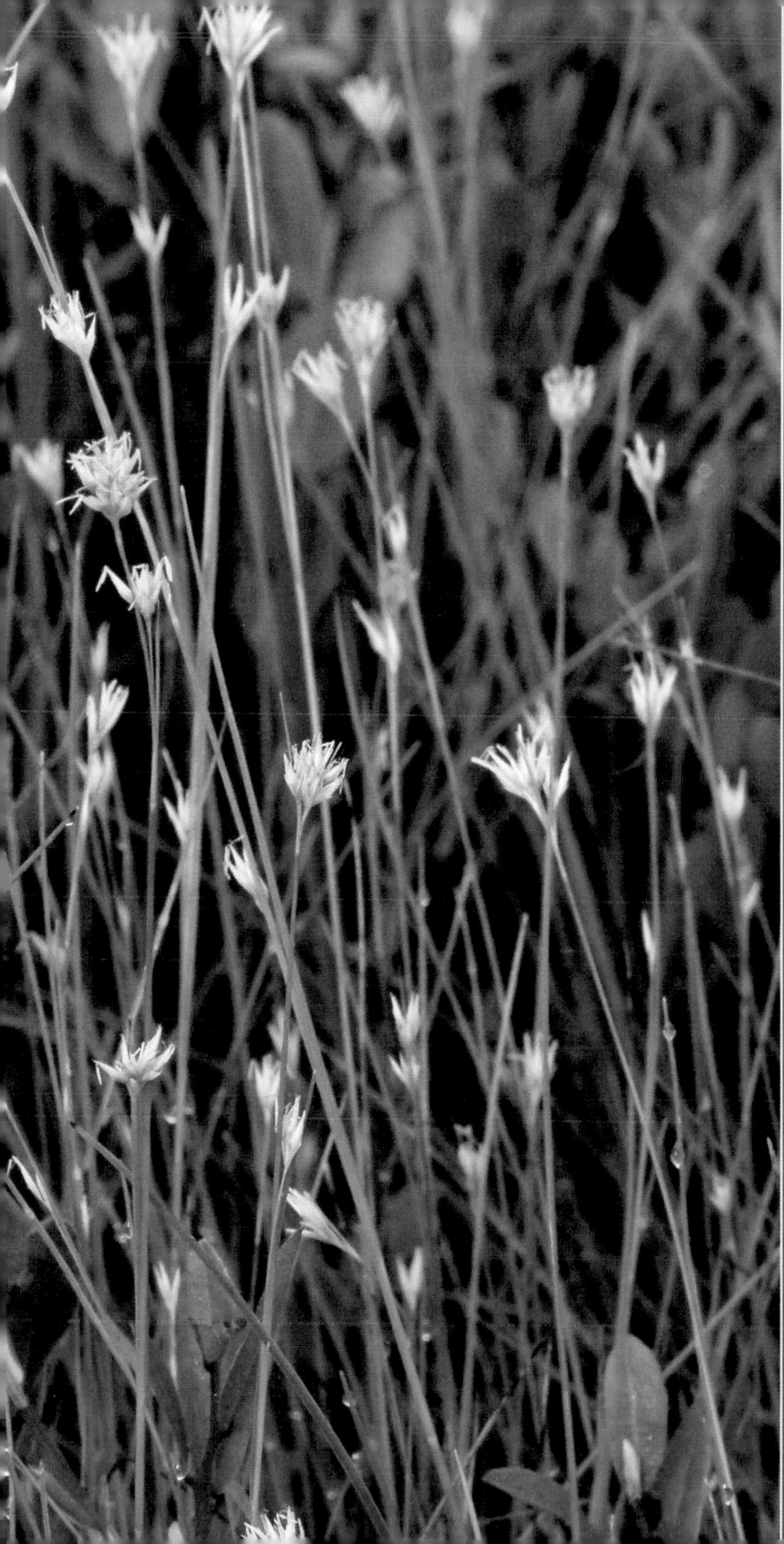

Herbaceous Plants

40

Saltmarsh arrowgrass Although its Latin name is *Triglochin maritima*, and it commonly occurs in marine saltmarshes, this species is included here because it also occurs in many of our inland fens. While its leaves superficially resemble those of grasses, as reflected by its common name, it is not a true grass in the family Poaceae. It is a perennial herb in the arrow-grass family, Juncaginaceae. This family is in the Alismatales, an order that includes several additional families with species of aquatic and wetland habitats, including Scheuchzeriaceae (see podgrass).

IDENTIFICATION The leaves are narrow, somewhat fleshy, arise from the base of the plant, and are sheathed at their base (40.1, 40.2). They are more or less erect, as long at 20 in. (0.5 m), and only 0.04–0.12 in. (1–3 mm) wide. The flowering stalk (scape) is 8–40 in. (0.2–1 m) tall, with the top 4–16 in. (0.1–0.4 m) bearing the greenish and red-brownish perfect flowers (40.1, 40.3). Each flower is borne on a short (0.2 in. or 5 mm) pedicel and has

six 0.08–0.2 in. (2–5 mm) oblong ovaries attached to a central axis. Each ovary has a stigma with radiating papillae (40.3). The fruit is a follicle (40.4). One other *Triglochin* species occurs in the fens of our region: marsh arrowgrass or slender bog arrowgrass, *Triglochin palustre* (or *palustris*) (OBL). It is readily distinguished from *T. maritima* by its smaller and slenderer build and by having only three pistils per flower.

40.1 Saltmarsh arrowgrass: plants in a fen, accompanied by sweetgale and other species. Photograph: Russ Schipper. **40.2** Saltmarsh arrowgrass: leaf bases showing two translucent sheaths. Photograph: Arthur Haines. **40.3** Saltmarsh arrowgrass: part of a flowering scape, showing papillose styles. Photograph: Robert W. Smith. **40.4** Saltmarsh arrowgrass: part of a fruiting scape, showing follicles with stygmatic beaks. Photograph: Russ Schipper.

OCCURRENCE · OBL *Triglochin maritima* occurs in all 19 of our states and provinces. It reaches its eastern North American southern limit in our southern tier of states from Illinois to New Jersey, except it occurs farther south in Maryland and Delaware. It is listed as endangered in New Jersey and threatened in Illinois and Ohio. *Triglochin palustre*'s distribution is similar, except it is not recorded for Vermont, Massachusetts, Connecticut, and New Jersey, nor does it occur in Maryland or Delaware. It is listed as threatened in Illinois, Indiana, New York, extirpated in Pennsylvania, and historical in Rhode Island.

Triglochin maritima occurs in circumneutral and alkaline fens and is generally more abundant in the fens of the western part of our area. It also occurs inland in meadows and marshes, at lake and stream shorelines and beaches, often associated with calcareous substrates. Additionally, and as its names indicate, it is common in marine tidal saltmarshes and is even found in brackish hollows at rocky marine shores. *Triglochin palustre* occurs in most of these same habitats.

41

Calico aster

or side-flowered aster (*Symphyotrichum lateriflorum* var. *tenuipes*) is a perennial herb in the aster family, Asteraceae. The species has been variously categorized and divided into varieties, only some of which occur in peatlands. Calico aster is one of several aster species that grow in our peatlands.

IDENTIFICATION This variety of calico aster grows 1–4 ft. (0.3–1.2 m) tall and has narrow, alternate leaves. The principal leaves are 2–6 in. (5–15 cm) long, but those on the side branches are much reduced (41.1, 41.2). Flower heads are distributed along the side branches, the terminal heads developing first (41.1, 41.2). There are 9 to 14 white rays per head (41.2, 41.3). A head in fruit appears like a ball of fuzz. Its multiple fruits are achenes, each one with numerous bristles that facilitate wind dispersal. Another member of this genus that commonly occurs in peatlands is rush American aster or northern bog aster, *S. boreale* (OBL). It has narrow linear leaves, many (commonly more than 20) white to pale blue or lavender rays per head, and these heads are not limited to side branches.

OCCURRENCE · FACW The nominate variety of calico aster, *S. lateriflorum* var. *lateriflorum*, occurs throughout our area but rarely in peatlands. *Symphyotrichum lateriflorum* var. *tenuipes* occurs at open and wooded sites at the periphery of fens and bogs, as well as at mineral-soil wetlands. It is more northeastern than the nominate variety and is absent from Wisconsin, Illinois, Indiana, Ohio, Pennsylvania, New Jersey, Connecticut, Rhode Island, and Massachusetts. *Symphyotrichum boreale* occurs in all of our area except Massachusetts, Connecticut, and Rhode Island but becomes rare

at the southern fringes of the area. It grows in open and open-wooded fens, often of the circumneutral to alkaline variety, and also in mineral-soil wetlands and along lake and river shorelines.

41.1 Calico aster: plant with flower buds. **41.2** Calico aster: flowering stem. **41.3** Calico aster: close-up of head, cropped from 41.2..

42

Baked-apple berry or cloudberry (*Rubus chamaemorus*) is a perennial herb in the rose family, Rosaceae. It is related to raspberry and blackberry but is spineless. The edible fruit is a favorite of boreal and Arctic peoples.

IDENTIFICATION The aboveground stems arise from creeping rhizomes. The flower stems are 4–10 in. (10–25 cm) tall. The leaves are round or kidney shaped and shallowly five- or seven-lobed (42.1–42.3). The largest leaves are 1.5–3.5 in. (4–9 cm) across. The white flowers are imperfect (unisexual), but they may have a few nonfunctional parts of the opposite sex. Flowers occur one to a plant, either pistillate (42.1) or staminate (42.2), and at the tip of the stem. The "berry" is an aggregate of small fruits, each one a single-seeded drupe. It is red tinged at first (42.3), then amber, then yellowing when fully ripe.

42.1

42.1 Baked-apple berry: plants, one with a pistillate flower. **42.2** Baked-apple berry: a plant with a staminate flower. **42.3** Baked-apple berry: plants with fruits that are aggregates of drupes. Photograph: Glen H. Mittelhauser.

OCCURRENCE · FACW This species occurs in the lowland peatlands of Québec and in the coastal peatlands of Nova Scotia, Prince Edward Island, New Brunswick, and eastern Maine. It also occurs in the peatlands of the Saint Lawrence valley of eastern Ontario. Elsewhere in the mapped part of Ontario (map 1) it is restricted to the peatlands of the far north. Additionally, it occurs at subalpine and alpine fens and peaty soils in the mountains of Maine, New Hampshire, and the Gaspé Peninsula of Québec. It is absent from Vermont, Massachusetts, Connecticut, Rhode Island, New Jersey, Pennsylvania, Ohio, Indiana, Illinois, Michigan, and Wisconsin.

43

White beaksedge (*Rhynchospora alba*) is a perennial herb in the sedge family, Cyperaceae.

IDENTIFICATION The plants grow in dense to loose groups (43.1). The flowering stems are 8–20 in. (0.2–0.5 m) tall, generally less than 14 in. (0.35 m) tall at ombrotrophic sites, and usually overtop the narrow leaves. The flowers are white, and each occurs in a crowded hemispheric cluster (43.2). The flowers increasingly tinge with tan as they age, and by late season produce tan, dry fruits (achenes).

OCCURRENCE · OBL This species is found at open areas of bogs and fens throughout our region. It is most abundant at wet moss lawns and the edges of pools.

43.2

43.1 White beaksedge: a dense stand of plants in flower.
43.2 White beaksedge: four flower clusters.

44

Devil's beggar-ticks or sticktights (*Bidens frondosa*) is an annual herb in the aster family, Asteraceae.

IDENTIFICATION This plant is erect (44.1) and generally 1–2 ft. (0.3–0.6 m) tall in peatlands, but can be at least 4 ft. (1.2 m) tall elsewhere. The leaves are compound, with three or five leaflets, but a few leaves may be simple (44.1). The terminal leaflets are lanceolate and up to 4 in. (10 cm) long, pointed, and serrate. Each flower head is shaped like an upside-down bell of variable proportions, and bears many small yellow-orange flowers (44.1). Ray flowers are absent in this species. The head is subtended by 5 to 10 (typically 7 to 8) green, leaflike bracts (44.1). Each flower produces a single-seeded dry brown fruit (an achene) with two barbed points (44.2), hence the name *Bidens*, meaning two teeth. These fruits stick to animals and, annoyingly, also to human clothing, and in this way the plant is dispersed.

OCCURRENCE · FACW This species occurs throughout our area at open and open-wooded sites at the periphery of fens and bogs, wet lake and river shorelines, ditches, and damp waste places. Because of its method of dispersal, it may be abundant along trails.

44.1 Devil's beggar-ticks: plant in flower. **44.2** Devil's beggar-ticks: a head in fruit, showing barbed teeth of the achenes.

45

Horned bladderwort or naked bladderwort (*Utricularia cornuta*) is a perennial herb in the bladderwort family, Lentibulariaceae. It is one of the "carnivorous" plant species that grows in our bogs and fens.

IDENTIFICATION When not flowering or fruiting, this plant is virtually invisible to the naturalist unless she or he knows exactly where to look and uses magnification. The near-colorless fine stems and threadlike leaves spread through the upper wet peat. The leaves have minute (about 0.06 in. or 1.5 mm) bladders that trap tiny invertebrates. The plant puts up straight flower stems 4–14 in. (10–35 cm) tall (45.1) that bear from one to three (rarely more) bright yellow bilaterally symmetrical flowers, each with an erect upper lip and a saclike lower lip that bears from its base a downward-projecting spur or "horn" (45.2). Each flower forms an egg-shaped fruit (a capsule) about 0.12 in. or 3 mm wide (45.3). Another bladderwort species that occurs in our peatlands is flat-leaved bladderwort, *U. intermedia* (OBL). It has dichotomously branched, finely divided, green leaves with flat ultimate divisions. These leaves typically are submerged in water but are also capable of creeping on wet surfaces when water level drops to expose bottom peat or sediment. Bladder size averages slightly larger than in *U. cornuta*. The yellow flowers are similar to those of *U. cornuta*, but the relative size of the upper lip is smaller, the large lower lip is nearly flat, and its spur is narrow and curves under it.

45.2

45.3

45.1 Horned bladderwort: a stand of plants in flower. **45.2** Horned bladderwort: close-up of a cluster of three flowers, with lateral view of flower on the right. **45.3** Horned bladderwort: close-up of a cluster of three fruits (capsules).

OCCURRENCE · OBL *Utricularia cornuta* occurs throughout our area on wet peats in bogs and fens, often where the peat surface has been disturbed, and occasionally at wet peaty, sandy, or muddy shores, boggy pools, and seeps. *Utricularia intermedia* occurs throughout our area in shallow water bodies associated with peatlands. These waters may be circumneutral or alkaline. It also occurs in shallow water of swamps, lakes, and streams.

46

Boneset or boneset thoroughwort (*Eupatorium perfoliatum*) is a perennial herb in the aster family, Asteraceae. It has been used for medicinal purposes but has potentially harmful side effects.

IDENTIFICATION This plant (46.1) is 2–5 ft. (0.6–1.5 m) tall, typically at the middle of that range in our peatlands. The stem is hairy (46.1). The 2.5-to-8-in.-long (6–20 cm) opposite leaves are serrate and gradually taper from their broad base to a pointed tip. The opposite leaves are fused (connate) at their base and surround the stem (46.1). On the basis of the "doctrine of signatures," which claims that an herb that resembles a part of the body can be used to treat ailments of that part of the body, this leaf fusion led to the belief that extracts of the leaves would heal broken bones, hence the name boneset. The inflorescence contains many heads, and each head (capitulum) contains from 9 to 23 white disk flowers (46.2). Ray flowers are absent. The fruits (achenes) have bristles that facilitate wind dispersal (46.3).

OCCURRENCE · FACW Boneset occurs throughout our area at open and open-wooded sites at the periphery of fens and bogs, and at swamps, wet thickets, marshes, lake and stream shorelines, and wet ditches.

46.1 Boneset: plant. **46.2** Boneset: inflorescence with open heads. Photograph: Donald S. Cameron. **46.3** Boneset: part of an inflorescence in fruit. Each near-spherical fruiting head has multiple achenes, each achene with outward pointing bristles. Photograph: Greg Vaclavek.

47

Buckbean or bogbean (*Menyanthes trifoliata*) is a perennial herb in the buckbean family, Menyanthaceae.

IDENTIFICATION The creeping stems of this species bear upright compound leaves 4–10 in. (10–25 cm) tall, each leaf with three leaflets at the top (47.1). The creeping stems also bear upright stalks with inflorescences on top containing white to pink five-petaled and very beautiful flowers (47.1). The petals are bearded on the upper surface (47.2). The fruits (capsules) mature from green to yellow to yellow tan (47.3) before they split open (47.4) to shed their shiny seeds.

OCCURRENCE · OBL This species occurs throughout our area at fens and the minerotrophic periphery of bogs in shallow pools and on wet surfaces, both at open and open-wooded sites. It also occurs at pond and lake margins.

47.1 Buckbean: plant in flower. **47.2** Buckbean: close-up of flower.
47.3 Buckbean: fruits (capsules). **47.4** Buckbean: split open capsules.
Photograph: Steve Matson.

47.1

48

Bunchberry, Canada dwarf-dogwood, or dwarf cornel (*Chamaepericlymenum canadense*, formerly *Cornus canadensis*), is a perennial herb in the dogwood family, Cornaceae.

IDENTIFICATION Bunchberry spreads by underground stems to form clonal colonies of plants. The plant is 2–9 in. (5–23 cm) tall and has two or three pairs of main opposite leaves, and one or two pairs of very small scalelike leaves below them. The main leaves seem whorled (all arising at same node) because the internodes are so short. In plants with three pairs of main leaves, the bottom pair is larger than the top two (48.1, 48.3). The leaf veins are arcuate and more or less parallel to the leaf edge (48.1, 48.3). An inflorescence that is subtended by four white bracts (modified leaves) is formed at the top of plants with three main leaf-pairs (48.1). The inflorescence contains tiny (about 0.07 in.; 1.5–2 mm) flowers (48.2), each forming a green drupe that turns red at maturity (48.3). These fruits are an important food for wildlife.

OCCURRENCE · FAC This species is widespread and common throughout our area at evergreen and mixed wooded sites in bogs and fens, and at non-wetland but moist evergreen and mixed forests, often mossy.

48.2

48.1 Bunchberry: plant in flower. **48.2** Bunchberry: close-up of an inflorescence. **48.3** Bunchberry: plants with fruits (drupes).

49

American bur-reed (*Sparganium americanum*) is a semi-aquatic and aquatic herb in the same family as cattail, Typhaceae.

IDENTIFICATION The ascending leaves (49.1) and floating strap-like leaves of this species can be as long as 36 in. (0.9 m). Numerous small unisexual flowers occur in separate globose heads, pistillate and staminate on the same plant (49.1, 49.2). The pistils have beak-like styles that give the head a bur-like appearance (49.2). The ovaries develop into dry fruits (achenes) that remain beaked.

OCCURRENCE · OBL This species occurs throughout our area at pools and running water associated with bogs, fens, and marshes, and at the shallows of fully aquatic systems.

49.2

49.1 American bur-reed: plant entering flower stage. The larger heads are pistillate.
49.2 American bur-reed: two staminate heads at on the left, one of them spent. The other three heads are pistillate. Photograph: Glen H. Mittelhauser.

50

Wild calla (*Calla palustris*) is an aquatic to semiaquatic perennial herb in the arum family, Araceae. The plant is poisonous owing to high oxalic acid content.

IDENTIFICATION The plant spreads by green horizontal stems at the water or wet soil surface. These stems bear upright lustrous leaves with heart-shaped blades. The blades are 2–4 in. (5–10 cm) long (50.1). Small bisexual (perfect) flowers, each with a green pistil and white anthers, are borne on a fleshy spadix. The spadix is subtended by a nearly flat, white spathe (50.1, 50.2). The uppermost flowers on the spadix may be imperfect and only staminate (50.2). The berries on the spadix develop from green to orange (50.3) and turn red at maturity (50.4).

OCCURRENCE · OBL Wild calla occurs throughout our area at shallow water or very wet areas of fens and similar areas at the minerotrophic periphery of bogs, as at the lagg. It also occurs at mineral-soil marshes, swamps, shallow ponds, and lake borders.

50.1 Wild calla: plants in flower. **50.2** Wild calla: close-up of flowers on spadix, subtended by white spathe. **50.3** Wild calla: developing fruits (berries). **50.4** Wild calla: ripe berries. Photograph: Louis M. Landry.

50.1

51

Canada mayflower or wild lily-of-the-valley (*Maianthemum canadense*) is a perennial herb in the family Ruscaceae (Haines 2011) but has recently been put in other families by other authors.

IDENTIFICATION This species has shiny parallel-veined and heart-shaped leaves. The plants emerge in the spring from underground rhizomes and have from one to three leaves per plant (51.1, 51.2). The plant is 1–8 in. (2.5–20 cm) tall. Groups of white, four-petaled flowers (51.3) are produced on two- or three-leaved plants (51.1, 51.2). The flowers produce green berries that become white and speckled with red, and if not eaten by wildlife will ultimately turn all red (51.4).

OCCURRENCE · FACU This species is common and widespread in non-wetland forests throughout our area. In peatlands, it occurs at minerotrophic, wooded, or open-wooded sites on hummocks above water level.

51.1

51.1 Canada mayflower: stand of plants. **51.2** Canada mayflower: a flowering plant.
51.3 Canada mayflower: close-up of inflorescence. **51.4** Canada mayflower: developing
and mature fruits (berries).

52

Tufted clubsedge

This species, *Trichophorum cespitosum*, is a sedge in the family Cyperaceae. It is an herbaceous perennial.

IDENTIFICATION Tufted clubsedge, as the name implies, grows in tufts or tussocks, that is, it is cespitose. The plants within a tussock form a clone (all the plants are genetically identical). In this species the tussocks are very dense (52.1). The individual stiff wiry stems are round-oval in cross section and 4–20 in. (0.1–0.5 m) long. Stems bear a "club" (a small solitary spikelet) at the top (52.2 to 52.5). The spikelet contains from three to nine perfect (bisexual) flowers and is covered by scales that are reddish to yellowish brown. The spikelets in one tussock display fresh stigmas or fresh stamens, not both at the same time. Photos 52.3 to 52.5 suggest that, at least in some tussocks in some geographic areas, the stamens are produced first (protandry), a condition that avoids self-fertilization within a tussock. The fruit is a hard achene.

Alpine clubsedge, *Trichophorum alpinum* (OBL), also occurs in our peatlands. As in *T. cespitosum*, it has tufts of stems with terminal spikelets, but differs in having long white bristles that emerge from the spikelets. Further, its stem is rough and has a triangular cross-section, rather than smooth and round or oval as in *T. cespitosum*.

52.1 Tufted clubsedge: tussocks in autumn. Photograph: Marilee Lovit. **52.2** Tufted clubsedge: young tussock with spikelets showing anthers, and with the prior year's dead stems. Photograph: Marilee Lovit. **52.3** Tufted clubsedge: spikelet with stamens. No other flower parts evident. Photograph: Marilee Lovit.

52.4 Tufted clubsedge: spikelets with groups of fresh trifid stigmas extending upward from each one. Wilting white filaments appear to have dropped their anthers. Some spent anthers (brown) remain. Photograph: Hermann Schachner. **52.5** Tufted clubsedge: two spikelets. The one in the foreground is closed. The one in the background bears trifid stigmas that are drying out at the tips. Photograph: Marilee Lovit.

OCCURRENCE · OBL Tufted clubsedge has a predominantly circumarctic and circumboreal distribution. Additionally, it has disjunct populations high on more southern mountains, as in the Appalachian and Rocky Mountains of the continental United States. It occurs in the jurisdictions of our area except Indiana, Ohio, Pennsylvania, New Jersey, Connecticut, Rhode Island, and Massachusetts. It is extremely rare in Vermont and is listed as threatened or endangered in Illinois, New York, and Wisconsin. In our area, it occurs in lowland fens and bogs, acidic to alkaline, and along lake and river shorelines. Also, it is commonly encountered in open and semi-open alpine and subalpine habitats, including moist pockets between rocks, as well as in alpine meadows. Alpine clubsedge has a similar geographic distribution, but in the peatlands of our area is largely restricted to circumneutral and alkaline fens. It can be common in such fens at the north-central and northwestern parts of our area. It also grows in meadows and fields and at lake and river shores, most commonly on calcareous substrates.

53

Tall cottonsedge (*Eriophorum angustifolium* ssp. *angustifolium*) is a perennial herb in the sedge family, Cyperaceae. In this book, I include three of the seven species of cottonsedges of our area. These three are the most common and widespread ones in our peatlands. See Gleason (1968) and Haines (2011) for species not included here.

IDENTIFICATION This genus has spikelets that, when mature, have numerous, elongate, and more or less straight bristles, in aggregate resembling a soft cotton brush. This species occurs in loose clumps of plants 12–30 in. (0.3–7.6 m) tall (53.1). Each plant has three to seven individual spikelets, each with numerous flowers. The spikelets are erect at first, later nodding as their stalks elongate (53.2 to 53.4) and the bristles develop. At least at some sites, the spikelets are initially pistillate (53.2) and later become staminate (53.3), a sequence called protogyny. Each flower produces a small (0.08–0.10 in.; 2–2.5 mm) dark brown fruit (an achene) surrounded by typically white bristles (53.1, 53.4), but the bristles may be pale yellow-brown in some populations. The group of spikelets is subtended by two to three leaflike bracts, one or two of which extend beyond the spikelets (53.1 to 53.4). The tips of leaves and bracts are often reddish, and the belowground stem and roots typically are pink.

OCCURRENCE · OBL This species occurs in all our states and provinces except Ohio, Pennsylvania, Massachusetts, Connecticut, and Rhode Island, and becomes rare at the southern fringes of its range in our area. Its range extends northward into the high Arctic. It occurs in open areas of fens and bogs, often at wet hollows and lawns and near pools.

53.4

53.1 Tall cottonsedge: group of plants in fruit. **53.2** Tall cottonsedge: plant in flower, with spikelets showing the three-cleft styles of the individual flowers. The developing bristles are barely visible in the bottom two spikelets. Photograph: Marilee Lovit. **53.3** Tall cottonsedge: plant in flower, with spikelets showing the anthers of the individual flowers. The anthers are underlain by the developing bristles. Photograph: Marilee Lovit. **53.4** Tall cottonsedge: close-up of spikelets in fruit. The small achenes are hidden near the bases of the bristles.

54

Tawny cottonsedge (*Eriophorum virginicum*) is a perennial herb in the sedge family, Cyperaceae. See tall cottonsedge for general information on cottonsedges.

IDENTIFICATION This species occurs in open stands of plants 20–43 in. (0.5–1.1 m) tall (54.1). Each plant has 2 to 10 spikelets in a crowded cluster, typically appearing as a single dense head (54.2, 54.3). The bristles are light tan to brown, at least at the base (54.1, 54.3), and rarely all white. Near their base they enclose approximately 0.13-in.-long (3–3.5 mm) fruits (achenes). Arising from the base of the group of spikelets are two to five leaflike bracts, the longest 2–5 in. (5–13 cm) (54.2, 54.3).

OCCURRENCE · OBL This species occurs at open and open-wooded areas of bogs, acidic fens, and meadows throughout our area. It is the most southern of the three cottonsedges in this book, occurring north of our area but not to the Arctic, and south of our area as far as Georgia.

54.1

54.1 Tawny cottonsedge: stand of plants. **54.2** Tawny cottonsedge: plant in flower. At least four closely packed spikelets are distinguishable. Photograph: Marilee Lovit. **54.3** Tawny cottonsedge: plant with dense aggregate of spikelets in fruit. The achenes are hidden near the bases of the bristles.

55

Tussock cottonsedge (*Eriophorum vaginatum* ssp. *spissum*) is a perennial herb in the sedge family, Cyperaceae. See tall cottonsedge for general information on cottonsedges.

IDENTIFICATION This species occurs in very crowded clumps (tussocks) of plants 7–24 in. (0.2–0.6 m) tall (55.1), each plant with only one erect and near-globose spikelet at the top of the stem. The bristles are white (55.1, 55.2), rarely reddish or brown, and conceal the approximately 0.12-in.-long (3 mm) fruits (achenes) near their base. Earlier in development, the flowering spikelet displays numerous gray, pointed, and more or less translucent scales (55.3), each subtending a separate flower. There are no leaflike bracts subtending the spikelet. The base of the plant under the peat surface has abundant brown fibrous sheaths.

OCCURRENCE · OBL Tussock cottonsedge occurs throughout our area except Indiana and Ohio. It grows at open sites at acidic fens and bogs, often on relatively dry upper peat. It has been observed to regenerate abundantly where peat has been disturbed or drained. It also occurs in alpine tundra on our highest mountains. This species has a predominantly boreal-Arctic distribution and is near its eastern North American southern limit in our area.

55.2

55.3

55.1 Tussock cottonsedge: fruiting plants densely packed in tussocks.
55.2 Tussock cottonsedge: close-up of fruiting plants in a small tussock.
The achenes are hidden near the bases of the bristles. 55.3 Tussock
cottonsedge: close-up of a flowering spikelet showing spent anthers
(yellow tan), pointed translucent scales, and underlying fine, partly
formed white bristles. Photograph: Marilee Lovit.

56

Greater water dock This species, *Rumex britannica*, formerly *R. orbiculatus*, is a large perennial herb in the buckwheat family, Polygonaceae.

IDENTIFICATION This herbaceous plant is 3–8 ft. (0.9–2.4 m) tall, including the inflorescence. The leaves are lanceolate, the basal ones 12–24 in. (0.3–0.6 m) long (56.1). The leaves along the fully developed stem are much smaller than the basal ones (56.2, 56.3). The inflorescence is tall, with many small green flowers (56.2). The flowers develop into green three-winged accessory fruits (56.3, 56.4), each consisting of an achene enveloped by the inner sepals. The fruits may turn pinkish green before becoming reddish brown and dry at maturity.

OCCURRENCE · OBL This widespread species occurs at peripheral areas of fens and bogs and also at swamps, wet meadows, both fresh and brackish marshes, shallow water, lake, river, and stream shorelines, and ditches. It occurs in all 19 of our states and provinces.

56.1 Greater water dock: plants with basal leaves only. **56.2** Greater water dock: top of plant in flower. **56.3** Greater water dock: top of plant in fruit. **56.4** Greater water dock: close-up of accessory fruits (see text).

57

Bog goldenrod This species of perennial herb, *Solidago uliginosa*, is the most common and widespread of the few species of goldenrods that grow in our peatlands. It is a member of the aster family, Asteraceae.

IDENTIFICATION Bog goldenrod is quite variable, and it may hybridize with other *Solidago* species, taking on intermediate characteristics. Its typically solitary and unbranched stems range greatly in height, from 1 to 4 and rarely to 6.5 ft. (0.3–1.2 and rarely to 2 m). The alternate leaves are elongate, ranging from as long as 14 in. (35 cm) at the base of the plant (57.1) to less than 1 in. or less than 2.5 cm long at the tip of the plant (57.2). Those arising from the stem near the base of the plant are sparsely serrate, with winged petioles that clasp halfway or more around the stem. Leaf blade size, serration, petiole length, and clasping are progressively reduced upward on the stem, with leaves at the upper half of the stem lacking serration, petioles, or clasping (57.1, 57.2). The inflorescence at the top of the stem is commonly long and near-cylindrical, typically not much more than two times wider near the base than near the tip (57.2, 57.3), but there are exceptions where the base is proportionately wider. The length of the inflorescence also varies widely. The yellow flower heads have five to eight disk flowers and one to eight rays. The fruits are achenes with numer-

ous bristles that facilitate wind dispersal.

OCCURRENCE · OBL Bog goldenrod occurs in all our states and provinces. It grows in fens and the minerotrophic edges of bogs, wet meadows, marshes, swamps, and river and lake shores, any of these habitats ranging from acidic to alkaline.

57.3

57.1 Bog goldenrod: Lower half of plant showing decreasing leaf size upward on stem. Photograph: Leanne Wallis. 57.2 Bog goldenrod: upper third or so of stem with small leaves and inflorescence. Photograph: Allen Chartier. 57.3 Bog goldenrod: most of an inflorescence. Photograph: Robert W. Smith.

58

Three-leaved goldthread or canker root (*Coptis trifolia*) is a perennial, evergreen herb in the buttercup family, Ranunculaceae. The bitter golden rhizomes are used as medicine.

IDENTIFICATION This low-to-the-ground plant spreads by bright golden rhizomes. Each shiny compound leaf has three leaflets, not three leaves as in the common name of the plant (58.1). The leaflets are shallowly lobed and serrate. Flowering stems are 2–5 in. (5–13 cm) tall, each bearing a single flower (58.2). The flowers lack petals but have 5 to 7 white petal-like sepals, 5 to 25 white stamens, and 3 to 7 yellow-and-green stalked pistils. Each pistil produces a follicle (58.3).

OCCURRENCE · FACW This species occurs at wooded sites at fens and the periphery of bogs; also at moist but non-wetland forests throughout our area, except Illinois.

58.1 Three-leaved goldthread: stand of plants showing leaf details. **58.2** Three-leaved goldthread: stand of plants in flower. **58.3** Three-leaved goldthread: stalk with four fruits (follicles).

59

Canada reed grass or bluejoint (*Calamagrostis canadensis*) is a perennial herb in the grass family, Poaceae.

IDENTIFICATION Grasses, while typically less plentiful than sedges in our peatlands, may be abundant in some fens, and at least are present in small numbers in some part of all, or almost all, of our peatlands. Grasses differ from sedges in having round, hollow stems with joints (nodes) where the leaves emerge, and leaves that alternate on the stem in approximately a single plane so that a cut plant lies nearly flat on the ground. See boreal bog sedge for distinguishing characteristics of sedges.

Canada reed grass occurs in clumps, 20–60 in. (0.5–1.5 m) tall. The nodes are often stained blue or purple. The leaves are overtopped by the flowering stems (culms) (59.1). Each culm has an inflorescence (panicle type) at the top consisting of many small spikelets (59.2). The inflorescence is conical when its spikelets are open (59.3). The spikelets typically are purple and white. The fruit is a small tan grain, 0.05–0.08 in. (1.25–2 mm) long.

OCCURRENCE · OBL This species occurs at wet, open, and open-wooded areas at the periphery of our peatlands and may be abundant and widespread at mineral-soil marshes and meadows. It is common in many wetlands in all 19 of our jurisdictions.

59.1 Canada reed grass: tops of plants in flower. **59.2** Canada reed grass: slightly nodding inflorescence (panicle). **59.3** Canada reed grass: upright, more open inflorescence (panicle).

59.1

60

Fowl manna grass

This species, *Glyceria striata*, is a perennial herb in the grass family, Poaceae. Where abundant, its grain (a type of fruit) is an important food for waterfowl and other birds.

IDENTIFICATION See Canada reed grass page for general characteristics of grasses. Fowl manna grass plants are 20–55 in. (0.5–1.4 m) tall and occur in clusters. The flat leaf blades are as wide as 0.4 in. (1 cm). Where a leaf blade joins the stem, it wraps around it to form a completely closed sheath (60.1). The lax and diffuse collection of fine branches in photo 60.1 is the inflorescence (a panicle). On these branches are numerous small light green spikelets. The spikelets in the photo are young and haven't yet expanded to reveal their flowers. Each spikelet, when mature, is about 0.15 in. (0.4 cm) long and displays its three to six small flowers (60.2). It may produce abundant grain, each grain being reddish and 0.02–0.08 in. (0.5–2 mm) long.

OCCURRENCE · OBL This species is very widespread in North America. It occurs throughout our area at open and open-wooded sites in marshes, swamps, lake and stream shorelines, fens, and the minerotrophic periphery of bogs.

60.1 Fowl manna grass: plant with panicle bearing young spikelets. The two white arrows indicate sheaths formed by leaf bases that wrap entirely around the stem. **60.2** Fowl manna grass: small part of panicle with flowering spikelets showing anthers. Photograph: Marilee Lovit.

61

Fen grass-of-Parnassus, or bog stars, *Parnassia glauca*,
is a perennial herb that has been variously placed in the saxifrage
family, Saxifragaceae; bittersweet family, Celastraceae; or its own
monogeneric family, Parnassiaceae, as followed by Haines (2011).
Despite its preferred common name, it looks nothing like a grass.

IDENTIFICATION This species has petiolate basal leaves with ovoid
blades 1–2.5 in. (about 2.6 cm) long. Rising from the base are
flower stalks 5–19 in. (about 12–50 cm) tall (61.1). A small leaf is
sometimes present as high as halfway up the flower stalk. Each
stalk is topped by a single gorgeous flower. The flowers have five
white petals 0.4–0.7 in. (1–1.8 cm) long, each with dark, typically
green veins. A three-part, sterile "staminodium" is present at the
base of each petal. The five larger regular stamens arise between
the petal bases (61.2). The fruit is a capsule.

OCCURRENCE · OBL The distribution of this eastern North American species is predominantly boreal but has been reported in all
19 of our states and provinces. It reaches its southern limit in our
southern tier of states from Illinois to New Jersey. In some of these
southern jurisdictions it is listed as rare, endangered, threatened,
or vulnerable. It is a plant of circumneutral and alkaline fens. As
such fens are more abundant in the western half or our area, it
occurs more regularly there than in the east. It also occurs on wet,
typically calcareous mineral soils at lake and stream shorelines
and seeps.

61.2

61.1 Fen grass-of-Parnassus: plants in bloom in a fen. Photograph: Anton A. Reznicek. 61.2 Fen grass-of-Parnassus: close-up of flowers. Photograph: Warren H. Wagner Jr., courtesy of University of Michigan Herbarium.

61.1

62

Blue iris or northern blue flag (*Iris versicolor*) is a strikingly beautiful perennial herb in the iris family, Iridaceae.

IDENTIFICATION The strap-like leaves are 15–35 in. (0.4–0.9 m) long (62.1, 62.2). The flowers are 2.5–4 in. (6–10 cm) wide, blue to violet with darker veins and areas of white and yellow (62.1 to 62.3). The capsular fruits are triangular in cross section, beaked (62.4), and dry out to a tan color by the end of the season.

OCCURRENCE · OBL Blue iris occurs throughout our area except Indiana. It grows at fens and minerotrophic peripheral areas of bogs, at both open sites and wooded edges. It also grows in mineral-soil marshes and wet meadows, at stream, river, and lake shores (62.1), and in drainage ditches.

62.2

62.1 Blue iris: plants in shallow lake water. **62.2** Blue iris: blooming plants at an open fen. **62.3** Blue iris: close-up of flower. **62.4** Blue iris: fruits (capsules), the one on the right starting to split open to reveal seeds.

63

Virginia marsh-Saint-John's-wort (*Triadenum virginicum*) is an herbaceous perennial in the Saint-John's-wort family, Hypericaceae.

IDENTIFICATION This species has a single erect stem 8–24 in. (0.2–0.6 m) tall that may be branched (63.1). The opposite ovoid leaves lack petioles and clasp the stem, the principal ones being 0.8–2.8 in. (2–7 cm) long. They are dotted with translucent glands and often have a reddish tint, at least along their smooth edges. The stem, too, may be red tinged (63.1, 63.4). Groups of flowers arise from the leaf axils and/or at the tips of stems (63.2). The five petals, each 0.25–0.4 in. (7–10 mm) long, are pinkish, reddish, or greenish. The nine stamens arise in three groups of three each, and the three-part ovary displays three styles (63.3). The fruit is a capsule, 0.3–0.5 in. (8–12 mm) long, typically red colored before completely drying (63.4) and consisting of three parts that split open when ripe to release the seeds. A more northern sister species, sometimes considered a variety or subspecies of *T. virginicum*, is Fraser's marsh-Saint-John's-wort, *T. fraseri* (OBL). The two taxa are very similar, *T. fraseri* having shorter petals, sepals, and styles and producing capsules that narrow to a point more abruptly.

63.1 Virginia marsh-Saint-John's-wort: plant. Photograph: Bruce Patterson.
63.2 Virginia marsh-Saint-John's-wort: stem tip in flower. Photograph: Marilee Lovit.
63.3 Virginia marsh-Saint-John's-wort: flower close-up. Photograph: Marilee Lovit.
63.4 Virginia marsh-Saint-John's-wort: capsular fruits.

63.3

OCCURRENCE · OBL *Triadenum virginicum* occurs in all 19 of our states and provinces, but some authorities recognize only *T. fraseri* for New Brunswick and Prince Edward Island. *Triadenum virginicum* is less abundant in the western part of our area and is rare and listed as endangered in Illinois. *Triadenum fraseri* is reported for all our jurisdictions except Rhode Island. Both species grow in a wide range of wetland habitats, including marshes and lake and stream shorelines and at peatland edges, but not in our most acidic peatland habitats.

64

Tall meadow-rue (*Thalictrum pubescens*) is very tall for a perennial herb in our area. It is in the buttercup family, Ranunculaceae.

IDENTIFICATION This plant grows to 8 ft. (2.4 m) tall and has compound leaves. The leaflets are typically three lobed but with some variation (64.1). Each lobe has smooth edges except for a short, fine point at the tip of the main lobe. Male and female plants are separate. The flowers occur in inflorescences that are compound and variable in shape and size (64.2 to 64.4). The typically white (may have some yellow or pink), staminate flowers (64.2, 64.3) have long, straight anthers and lack pistils. The white-to-pink-to-green pistillate flowers (64.4) may have stamens with sterile pollen. The pistils produce fruits of the achene type (64.4).

OCCURRENCE · FACW The species occurs throughout our area except Wisconsin. It grows at wet peripheral areas of fens and bogs and at swamps, marshes, meadows, fields, and lake, river, and stream shorelines.

64.1 Tall meadow-rue: leaves. **64.2** Tall meadow-rue: stem with staminate inflorescence. **64.3** Tall meadow-rue: close-up of part of a compound staminate inflorescence. **64.4** Tall meadow-rue: two greenish-white pistillate flowers with presumably sterile anthers; and clusters of young green-and-rose-colored achenes, each cluster from one flower. Photograph: Glen H. Mittelhauser.

64.1

65

White-fringed bog-orchid
This species, *Platanthera blephariglottis* var. *blephariglottis*, is a perennial herb in the Orchidaceae.

IDENTIFICATION The plant has one erect stem 8–30 in. (0.2–0.75 m) tall, with elongate alternate leaves. An inflorescence 2–7 in. (5–18 cm) tall by 2–3 in. (5–7.5 cm) wide of fragrant white flowers forms at the top (65.1, 65.2). The flower has a fringed lip and a curved, hollow spur 0.4–1.2 in. long (1–3 cm) descending from the underside of its base (65.1 to 65.3). The spur contains nectar that is sought by pollinators, especially moths with a long "tongue" (proboscis). In the process, the moth picks up masses of pollen from the pairs of yellow pollen sacs (65.3) and carries the pollen to a stigma (between the pollen sacs) of another flower. Fertilized flowers produce capsules that are green at first but turn tan and dry at maturity (65.4). These fruits split open to release large numbers of minute, wind-dispersed seeds.

OCCURRENCE · OBL Except for Wisconsin and Indiana, this species occurs in all the states and provinces of our area. It grows at open and open-wooded sites at fens and at minerotrophic and transitional minero- to ombrotrophic areas of bogs. It is listed as endangered, threatened, or vulnerable in several of our states.

65.1 White-fringed bog-orchid: plant starting to bloom. **65.2** White-fringed bog-orchid: inflorescence. **65.3** White-fringed bog-orchid: close-up of flowers. Front view of lower flower shows opening at the top of the lip to the nectar-containing spur. **65.4** White-fringed bog-orchid: ripe fruits (capsules) in the fall.

66

Dragon's mouth or dragon's-mouth orchid (*Arethusa bulbosa*) is a perennial herb in the Orchidaceae.

IDENTIFICATION The plant is 4–12 in. (10–30 cm) tall and consists of a single stem arising from an underground bulb. The stem has bractlike leaves and one elongate grasslike leaf, all arising from its lower half. A single flower, 1.5–2.0 in. (4–5 cm) tall is borne at the top of the stem (66.1, 66.2). The flower has lance-shaped, pink to magenta sepals and petals that resemble each other, point more or less upward, sometimes arched, and are united at the base (66.1, 66.2). The slightly convex lip projects forward, drooping somewhat, and has a wavy edge (66.1, 66.2). The lip is patterned white and magenta to pink (66.2) and may have yellow ornamentation toward its base (66.1). The fruit (66.3) is a 0.7–1.4 in. (1.8–3.5 cm) long-ellipsoid ridged capsule with a beak of variable length, green when young, then drying and browning before splitting open to shed its tiny seeds.

OCCURRENCE · OBL This species is found at wooded and open sites in bogs, fens, and wet meadows and often is associated with peat moss. It occurs in all the states and provinces of our area and is rare and threatened or endangered in all the states except Maine. In Maine, it occurs in many bogs and acidic fens, but rarely in abundance.

66.1 Dragon's mouth: frontal-oblique view of flower. **66.2** Dragon's mouth: front view of flower. **66.3** Dragon's mouth: fruit (capsule) with a longer beak than most. Photograph: Ellis Squires.

67

Grass pink, tuberous grass-pink, or grass-pink orchid (*Calopogon tuberosus*) is a perennial herb in the Orchidaceae.

IDENTIFICATION This species is 8–20 in. (0.2–0.5 m) tall in our area, with a narrow leaf at the base. The plant is inconspicuous when not in flower or fruit. It overwinters as a corm (solid bulb) in the upper peat. The flowers are pink or lavender or magenta (67.1, 67.2), sometimes white, and one to several per plant (67.1). The petals and sepals are separate and distinct, and spreading. Unlike most other orchids, it has flowers with the lip on top (67.1, 67.2). The lip is two-winged and bears a yellow-white brush (67.1, 67.2). The fruit (capsule) is green when young (67.3) but dries and turns brown at maturity.

OCCURRENCE · OBL Grass pink occurs in peatlands throughout our area, where it is the most common and widespread orchid species. It is found at open areas of sphagnous bogs and fens, in profusion at some sites.

67.1 Grass pink: plants in flower. **67.2** Grass pink: close-up of flower.
67.3 Grass pink: fruits (capsules).

Rose pogonia, or snake-mouth orchid (*Pogonia ophioglossoides*), is a perennial herb in the Orchidaceae.

IDENTIFICATION The aboveground part of this plant consists of a single stem 6–16 in. (15–40 cm) tall, typically 7–11 in. (18–28 cm) in our area. At mid-height the stem bears a large leaf (68.1). The single rose-pink to pale pink flower atop the stem is subtended by a small bractlike leaf (68.1 to 68.3). The flower is rarely white. It has separate and spreading sepals and petals and a fringed and bearded lip that widens out from the base (68.2, 68.3). The lip is marked with magenta lines; the hairs near its base are yellow (68.2, 68.3). The fruit (a capsule) is ridged and about 1 in. (2.5 cm) long, green at first (68.4) and drying to tan brown at maturity.

OCCURRENCE · OBL This species occurs at open and wooded sites in fens, sandy wet meadows, and ditches throughout our area. In bogs, commonly it occurs at acidic, minerotrophic, open and wooded sites, and at the transitional zone between minerotrophic and ombrotrophic areas.

68.1 Rose pogonia: three plants in flower. **68.2** Rose pogonia: close-up of flower in profile. **68.3** Rose pogonia: close-up of front of flower. **68.4** Rose pogonia: fruit (capsule). Photograph: Louis M. Landry.

69

Purple pitcher plant (*Sarracenia purpurea*) is a perennial herb in the pitcher plant family, Sarraceniaceae. It is one of a few kinds of carnivorous plants in our peatlands.

IDENTIFICATION This beautiful and distinctive plant has leaves 6–12 in. (15–30 cm) long, including a narrow base that may be partly underground. These leaves are shaped like pitchers with a hood on top (69.1). Typically, several pitchers radiate out from the growing tip of a subsurface stem (69.2). The pitchers collect rainwater. Insects and other small creatures are attracted to the pitchers by color (mimics flowers), scent, and nectar from glands near the top of the pitchers. The slippery interior and stiff downward-pointing hairs on the hood make escape difficult. Eventually, the struggling insect falls into the water and drowns. It is consumed and decomposed by insect larvae and microbes living in the pitcher. This process releases essential mineral nutrients like phosphorus that are used by the plant to supplement the meager supply of mineral nutrients in the peat. This process is not the same as carnivory in animals, because the plants do not derive energy from their prey. The plants have chlorophyll, and get their energy from photosynthesis. The ovary of the spectacular purplish-red and green flower

(69.2) develops into a dry fruit (69.3) akin to a capsule. It splits open to release its many seeds.

See http://www.sarracenia.com/faq/faq5520.html or http://www.botany.org/Carnivorous_Plants/Sarracenia.php for more details.

OCCURRENCE · OBL This species occurs at open and wooded sites in sphagnous bogs, acidic fens, and peaty shores throughout our area.

69.1 Purple pitcher plant: close-up of pitchers. **69.2** Purple pitcher plant: full plant in flower. **69.3** Purple pitcher plant: capsular fruit and shriveling flower parts.

70

Podgrass or rannoch rush (*Scheuchzeria palustris*) is an herbaceous perennial plant. It isn't really a grass or a rush and is the only species in the family Scheuchzeriaceae. This family is in the order Alismatales that includes several additional families with species of aquatic and wetland habitats, including Juncaginaceae (see saltmarsh arrowgrass).

IDENTIFICATION Podgrass plants arise from creeping rhizomes and are 8–16 in. (20–40 cm) tall. They have zigzag stems with alternate leaves whose apical portions are tubular, 0.04–0.12 in. (1–3 mm) wide, each with a blunt tip with a pore that is diagnostic of the species. Each leaf has a sheath where it joins the stem (70.1, 70.3), with the basal leaves having the longer and wider sheaths. The plant bears a terminal inflorescence 1.2–4 in. (3–10 cm) long and with several flowers (70.1, 70.2). The fruit is an ovoid, one- or two-seeded follicle, 0.25–0.4 in. (6–10 mm) long that matures from green (70.3) to red to brown (70.4) before opening to shed its seed.

OCCURRENCE · OBL Podgrass occurs throughout our area in sphagnous peatlands, peaty lake margins, and marshes. Its peatland habitats typically are wet and open areas of fens, very acidic to circumneutral, and typically in company with sedges.

70.1 Podgrass: the interior of a stand of plants showing one plant with an inflorescence. Photograph: George W. Hartwell, courtesy of Nancy Perkins. 70.2 Podgrass: close-up of inflorescence with a few flowers, displaying red-purple anthers and white styles with multiple tiny projections. Photograph: George W. Hartwell, courtesy of Nancy Perkins.
70.3 Podgrass: young follicles. Arrows indicate sheaths where leaves arise from the stem. Photograph: Glen H. Mittelhauser.
70.4 Podgrass: mature follicles splitting open, one of them showing a seed. Photograph: Glen H. Mittelhauser.

70.1

70.2

70.3

71

Wild sarsaparilla (*Aralia nudicaulis*) is in the carrot family, Apiaceae. It is a perennial that is commonly called an herbaceous plant, and I place it in that category here. But it has a woody stem barely or as much as 2 in. (5 cm) aboveground (71.1) that drops its single leaf in autumn, so it may legitimately be called a deciduous dwarf shrub. The stem, both the aboveground and belowground parts, is mildly aromatic and has been used as medicine and to flavor drinks. However, this species is not the sarsaparilla of commerce.

IDENTIFICATION The stiff petiole of the double-compound leaf rises from the woody stem and is divided at its top into three parts, each division bearing five (rarely more) leaflets (71.1). This unusual leaf form is diagnostic of the species. In some plants, a separate stalk rises from the stem and bears a compound inflorescence consisting of two to seven (commonly two to four) spherical units with white and green flowers (71.1, 71.2). Each flower produces a green fruit (berry) that matures to dark purple and retains the five styles of the pistil (71.3).

OCCURRENCE · FACU This species is found throughout our area on hummocks at wooded fens and the wooded, minerotrophic periphery of bogs. It is more common and widespread at non-wetland forests and forest edges.

71.1 Wild sarsaparilla: plant in leaf and in flower. White arrow points to a short woody stem. **71.2** Wild sarsaparilla: close-up of a spherical unit of an inflorescence. **71.3** Wild sarsaparilla: close-up of fruiting unit of an inflorescence.

72

Smooth saw-sedge (*Cladium mariscoides*) is an herbaceous perennial with the often-used but misleading common name twig-rush. The term "saw" in the common name used here is derived from another member of the genus that has sawtooth leaf edges, the southern (outside our area) saw grass, *Cladium mariscus* ssp. *jamaicense*. Here is a good lesson on the misleading aspects of common names. Neither of these species is a grass or a rush, nor does *C. mariscoides* have leaves with sawtooth edges. Both species are sedges in the family Cyperaceae. Some diagnostic characteristics of that family are given in the description of boreal bog sedge.

IDENTIFICATION This coarse sedge has stiff flowering stems that are 16–39 in. (0.4–1.0 m) tall and hard at the base (72.1). The leaves are only 0.04–0.14 in. (1–3.5 mm) wide and become round in cross section near the tip. The stem is topped by a branched inflorescence, 2–4 in. (5–10 cm) high (72.2). Smaller inflorescences may be present farther down the stem. Each inflorescence contains clusters of spikelets with reddish-brown scales (72.2, 72.3), often giving the inflorescence a golden to brown appearance (72.1 to 72.3). Photos 72.2 and 72.3 suggest, at least for these plants at this site, that the flowers are protandrous (stamens mature prior to the pistil), a mechanism that would prevent self-fertilization. The fruit is an achene.

72.1 Smooth saw-sedge: a stand of the species in a fen. Photograph: Russ Schipper.
72.2 Smooth saw-sedge: inflorescences in flower. The bright yellow projections
from most of the spikelets are anthers. The white projections from the out-of-focus
spikelets on the right are styles. Photograph: Russ Schipper. 72.3 Smooth saw-sedge:
spikelets showing trifid styles and spent anthers. Photograph: Russ Schipper.

OCCURRENCE · OBL This sedge occurs in all of our 19 states and
provinces. It is primarily a species of shallow freshwater of ponds
and lakes but may also occur in brackish marshes. It also occurs
along stream and river shores. It is common at the wet marshy
edges of our circumneutral and alkaline fens and at wet bog edges
in calcareous terrain, particularly in the central and western parts
of our area.

73

Boreal bog sedge (*Carex magellanica* ssp. *irrigua*) is a perennial herb in the sedge family, Cyperaceae. Some diagnostic characteristics of that family are given below.

IDENTIFICATION Sedges superficially look like grasses, but the two families are not closely related. Most sedges have triangular stems with keeled leaves arising alternately along successive angles of the stem. However, cottonsedge and some other sedge stems may not always appear triangular nor their leaves keeled. All sedge flowers are subtended by only a single scale. In the genus *Carex* (most peatland sedge species belong to this genus), the flowers are unisexual, and each pistillate flower is enclosed in a saclike scale called a *perigynium*, where the fruit (an achene) is formed. *Carex* flowers occur in spikes.

Boreal bog sedge forms loose clusters of plants 10–28 in. (25–71 cm) tall. The green leaves are 0.04–0.16 in. (1–4 mm) wide and have involute margins. The inflorescence consists of two to five pistillate spikes (73.1, 73.2), in some plants terminating in a staminate spike about 1 in. (2.5 cm) long (73.2). The lowest bract subtending the inflorescence is long and leaflike and extends beyond the inflorescence. Pistillate spikes are borne on drooping stalks, are 0.3–1 in. (0.8–2.5 cm) long, and have from 5 to 20 perigynia (73.1, 73.2). Each perigynium is subtended by a pointed scale that contrasts in color, is narrower than the perigynium, and typically exceeds the perigynium length (73.2).

73.1 Boreal bog sedge: plant in flower. **73.2** Boreal bog sedge: close-up of inflorescence with four spikes, three pistillate and one staminate.

The similar and closely related mud sedge (*Carex limosa*—OBL) averages about 4 in. (10 cm) shorter than boreal bog sedge. Also, in mud sedge, the lowest bract subtending the inflorescence is shorter than the inflorescence. Another difference from boreal bog sedge is that each perigynium is subtended by a scale that is shorter than, and about as wide as, the perigynium. As in boreal bog sedge, the scale is a contrasting color to the perigynium. Mud sedge also differs from boreal bog sedge in possessing foliage that is distinctly grayish or bluish green.

OCCURRENCE · OBL Boreal bog sedge occurs throughout our area except Illinois and Rhode Island in fens and weakly minerotrophic parts of bogs at well-vegetated open sites and at openings of wooded sites. Mud sedge occurs in all the states and provinces of our area. It grows at open and very wet sites in fens and bogs. It is a "pioneer species," often colonizing an invading peat mat at water edge or a wet exposed peat surface ("mud bottom") that has resulted from natural or anthropogenic disturbance.

Few-flowered sedge (*Carex pauciflora*) is an herbaceous perennial in the sedge family, Cyperaceae. It is quite distinctive and easily identified but can be inconspicuous and may be overlooked. The pointed perigynia stick to clothing and may be dispersed by animals.

IDENTIFICATION See the boreal bog sedge page for general features of sedges and the genus *Carex* and for definitions of some of the terms that follow. Few-flowered sedge is less than 16 in. (40 cm) tall and typically 8–12 in. (20–30 cm) tall. The stems are solitary or loosely clumped. The plant's one to three foliage leaves, when present, are much shorter than the stem and only 0.04–0.08 in. (1–2 mm) wide. The unique terminal spike bears from one to seven more or less reflexed (sharply bent downward), very narrow, and pointed perigynia, each 0.2–0.3 in. (5–8 mm) long (74.1). Two to four staminate flowers are borne in a slender terminal "cone."

OCCURRENCE · OBL Few-flowered sedge occurs in all our jurisdictions except Rhode Island, New Jersey, Ohio, and Illinois. It is at its southern limit in our area and becomes increasingly rare toward that limit. This species grows at open and partially shaded sites of bogs and acidic fens, often on *Sphagnum* mats.

74.1 Few-flowered sedge: terminal spike with reflexed perigynia. A stump of the staminate terminal "cone" remains at the tip. Photograph: Glen H. Mittelhauser.

75

Few-seeded sedge (*Carex oligosperma*) is an herbaceous perennial in the sedge family, Cyperaceae.

IDENTIFICATION See the boreal bog sedge page for general features of sedges and the genus *Carex* and for definitions of some of the terms that follow. Few-seeded sedge has stems up to 35 in. (90 cm) tall that occur in loose tufts (75.1). The long, wiry involute leaves are only 0.02–0.10 in. (0.5–2.5 mm) wide and have reddish-purple basal sheaths. There are one or two pistillate spikes per stem, each 0.4–0.8 in. (1–2 cm) long and with a long subtending bract (75.2, 75.3). A spike may contain from 3 to 25 perigynia, but commonly 9 to 15. The shiny perigynia are 0.16–0.27 in. (4–7 mm) long, beaked, inflated ovoids (75.3), each containing an achene when mature. The stem terminates in one or two staminate spikes (75.2).

OCCURRENCE · OBL This species occurs in all our states and provinces, with the possible exception of New Jersey. It is at its southern limit in our area, with the exception of rare occurrences in West Virginia and North Carolina, and becomes increasingly rare in our southern tier of states. At our mid-latitudes and northward it is among the most widespread and common sedge species at open sites in acidic fens and minerotrophic margins of bogs. It also occurs at lake and stream shorelines and wet meadows, where it may form extensive stands.

75.1

75.1 Few-seeded sedge: tufts of plants. Photograph: Glen H. Mittelhauser.
75.2 Few-seeded sedge: one pistillate and two staminate spikes. Photograph:
Glen H. Mittelhauser. 75.3 Few-seeded sedge: close-up of a pistillate spike
with perigynia. Photograph: Glen H. Mittelhauser.

76

Fringed sedge or drooping sedge (*Carex crinita*) is a perennial herb in the sedge family, Cyperaceae. It is a variable species that has been split into varieties or even separate species by various authors.

IDENTIFICATION See the boreal bog sedge page for general features of sedges and the genus *Carex* and for definitions of some of the terms that follow. Fringed sedge is 20–45 in. (0.5–1.1 m) tall and grows as a cluster of stems. It has leaves as broad as 0.5 in. (13 mm). The spikes are 1.5–4 in. (4–10 cm) long and become pendulous as they mature. The pistillate spikes appear fringed because of the projecting point of the scale that subtends each of its many perigynia (76.1).

OCCURRENCE · OBL This species occurs throughout our area at the wet edges of peatlands, often where there is shallow water. It also is found along lake and stream shorelines, in mineral-soil marshes, open swamps, wet fields, and ditches.

76.1 Fringed sedge: bottom and center, pistillate spikes; top and right, staminate spikes. Photograph: Glen H. Mittelhauser.

76.1

77

Hoary sedge or silvery sedge (*Carex canescens*) is a perennial herb in the sedge family, Cyperaceae.

IDENTIFICATION See the boreal bog sedge page for general features of sedges and the genus *Carex* and for definitions of some of the terms that follow. *Carex canescens* is 12–25 in. (0.3–0.6 m) tall and grows in a cluster of closely spaced stems (77.1). The plant is covered by a fine waxy bloom, giving it a slight silvery sheen. It has 3 to 8 spikes per stem (77.1), each spike with 5 to 30 flowers (predominantly 10 to 20) (77.2). Staminate flowers occur at the base of spikes below the perigynia. Each pistillate flower is enclosed in a white to green perigynium (77.2) that becomes yellowish brown at fruiting stage. The leaf blades are 0.08–0.16 in. (2–4 mm) wide.

OCCURRENCE · OBL Hoary sedge occurs throughout our area at wet places, usually with mosses, such as at fens and the minerotrophic periphery of bogs, and at both open and wooded sites.

77.1 Hoary sedge: plant tops in flower. **77.2** Hoary sedge: spikes showing styles emerging from perigynia. Stamens are visible on some spikes.

77.1

78

Meager sedge or coast sedge (*Carex exilis*) is an herbaceous perennial in the sedge family, Cyperaceae.

IDENTIFICATION See the boreal bog sedge page for general features of sedges and the genus *Carex* and for definitions of some of the terms that follow. Meager sedge occurs in dense tussocks, brown at the base, and has stiff leaves only 0.016–0.06 in. (0.4–1.5 mm) wide. The leaves are overtopped by sub-rigid flowering stems (culms) 6–26 in. (15–66 cm) tall, typically 10–16 in. (25–41 cm) tall (78.1). The spikes typically have a terminal pistillate part with spreading perigynia, and a staminate part below it (78.2 to 78.4). Two variations in a minority of plants are unisexual spikes, and typical spikes but with an additional staminate part at the tip. A similar and closely related species is dioecious sedge (*C. sterilis*—OBL). It differs from meager sedge in having two or more spikes per stem and in having pistillate and staminate spikes on separate plants with occasional exceptions. Like meager sedge, the plants grow in tussocks and have narrow stiff leaves, but the leaves are about twice as wide as those of meager sedge.

OCCURRENCE · OBL *Carex exilis* grows throughout our area except for Illinois, Indiana, and Ohio. Its occurrence is sporadic, but it can be locally abundant. It occurs at open and open-wooded (typically coniferous) peatlands, wet peat moss lawns of bogs and fens, meadows and fields, and along lake and river shorelines. *Carex sterilis* grows throughout our area in circumneutral and calcareous wetlands including fens, lake and river shorelines, meadows and fields, and swamp openings. It is rare or uncommon in northern New England and the Canadian Maritimes but common in the central and western parts of our area where wetland habitats are more often circumneutral or alkaline.

78.1 Meager sedge: tussock of plants. **78.2** Meager sedge: stems with terminal spikes. **78.3** Meager sedge: tops of two stems (culms) showing spikes. The terminal pistillate part has spreading perigynia; the scaly looking part of the stem below the perigynia is the staminate part. **78.4** Meager sedge: close-up of an older spike with more abundant perigynia than on the plants in photos 1–3. Photograph: Glen H. Mittelhauser.

79

Three-seeded sedge This species, *Carex trisperma*, is a perennial herb in the sedge family, Cyperaceae.

IDENTIFICATION See the boreal bog sedge page for general features of sedges and the genus *Carex* and for definitions of some of the terms that follow. Three-seeded sedge has weak, wirelike arching stems 5–26 in. (13–66 cm) long, and leaves only 0.03–0.08 in. (0.8–2.0 mm) wide (79.1). There are two to four spikes per stem, the lowest widely separated from the upper ones and subtended by a bristlelike bract. Each spike has only one to five perigynia (79.2). This sedge spreads by horizontal stems at ground level to form loose clumps of slender plants.

OCCURRENCE · OBL *Carex trisperma* is commonly present at fully or near-fully shaded sites in bogs and fens, and at wet woods with mixed or coniferous tree cover. It is especially abundant under black spruce. It occurs in all the states and provinces of our area.

79.2

79.1 Three-seeded sedge: clump of plants showing very slender stems and leaves.
79.2 Three-seeded sedge: inflorescences of two plants showing widely separated spikes with few perigynia. Photograph: Donald S. Cameron.

80

Woolly-fruited sedge or slender sedge (*Carex lasiocarpa* ssp. *americana*) is an herbaceous perennial in the sedge family, Cyperaceae.

IDENTIFICATION See the boreal bog sedge page for general features of sedges and the genus *Carex* and for definitions of some of the terms that follow. Woolly-fruited sedge is colonial and may form extensive stands. It grows 20–47 in. (0.5–1.2 m) tall. The plant has a slight waxy coating, and its basal sheath is reddish purple and fibrous. The leaves are long and narrow, typically less than 0.08 in. (2 mm) wide, and have prolonged threadlike apices (80.1). They are involute, with a U-shaped cross section. There are one to three cylindrical, 0.4-to-1.5-in.-long (1–4 cm) pistillate spikes per stem, each subtended by a leaflike scale (80.2). The perigynia are densely pubescent, hence the plant's common name, and each perigynium is subtended by a small purplish-brown scale (80.3). One to three staminate spikes, varying from 0.4 to 3.5 in. (1–9 cm) long, terminate the stem (80.2).

OCCURRENCE · OBL Woolly-fruited sedge occurs in all our states and provinces at wet open sites at minerotrophic edges of bogs, and at fens, especially at sites bordering open water. It also occurs in marshes and at lake, river, and steam shorelines and in shallow water, at some sites extending out from the shore to form a floating mat.

80.1 Woolly-fruited sedge: stand of plants. Note fine filamentous tips of leaves. Photograph: Glen H. Mittelhauser. **80.2** Woolly-fruited sedge: plant with two pistillate spikes and two terminal staminate spikes. Photograph: Marilee Lovit. **80.3** Woolly-fruited sedge: close-up of a pistillate spike, showing pubescent perigynia with feathery styles. Photograph: Marilee Lovit.

80.1

80.2

80.3

81

Skunk cabbage (*Symplocarpus foetidus*), a perennial herb in the arum family, Araceae, is the earliest herb to flower in the spring in our peatlands.

IDENTIFICATION The phenology or seasonality of skunk cabbage varies geographically depending on the climate, and especially the severity and duration of winters. The seasonal sequence described here and in the photos refers to central Maine. In late winter the spathes emerge from the surface by melting their way through frozen soil, ice, and/or snow by producing their own heat (81.1). By early spring they are fully grown and 3–6 in. (7.5–15 cm) tall (81.2). Each spathe encloses an ovoid, near-white spadix approximately 1 in. (2.5 cm) tall, and with many tiny flowers (81.3). Leaf buds open, and leaves start enlarging during the flowering period (81.2, 81.4), and they expand rapidly (81.5, 81.6). When fully

81.7

expanded the petiolate leaves have heart-shaped blades 8–20 in. (0.2–0.5 m) long. The leaves degrade during summer, and by late summer have withered away; but darkened and enlarged (to about 2 in. or 5 cm) spadices with embedded berrylike fruits persist (81.7). The plant is malodorous, as its name suggests.

OCCURRENCE · OBL Skunk cabbage occurs in minerotrophic parts of peatlands and mineral-soil wetlands in all 19 of our states and provinces, but it becomes less common in the northernmost parts of our area, where it is near its northern limit.

81.1 Skunk cabbage: overwintering leaf bud (gray), and a spathe (reddish) emerging through ice cover. Early March in central Maine. **81.2** Skunk cabbage: fully grown spathes in early April in central Maine. Emerging leaf tip visible. **81.3** Skunk cabbage: cutaway of a spathe showing spadix with tiny flowers. Early April in central Maine. **81.4** Skunk cabbage: plant with emerging leaves. Early April in central Maine.

81.5 Skunk cabbage: plant with young expanding and unfurling leaves. Late April in central Maine. **81.6** Skunk cabbage: plant with leaves about two-thirds expanded. Early May in central Maine. **81.7** Skunk cabbage: expanded spadix as a multiple fruit. Berrylike fruits are embedded in the spadix. September in central Maine.

82

Three-leaved false Solomon's seal (*Maianthemum trifolium*) is a perennial herb in the family Ruscaceae (Haines 2011) but has recently been put in other families by other authors.

IDENTIFICATION This species has alternate shiny, parallel-veined, and elongate leaves (82.1, 82.2), two to four per plant. The plant is 4–13 in. (10–33 cm) tall. It produces a group of white, six-petaled flowers (82.1). The flowers produce green berries speckled with red (82.2), which if not eaten by wildlife will ultimately turn all dark red.

OCCURRENCE · OBL This species occurs throughout our area except for Illinois and Indiana. It grows at wooded and open sites in bogs, fens, meadows, and moist woods. In bogs, it commonly occurs on a peat moss mat at the transition between a minerotrophic black-spruce wooded area and a more open ombrotrophic area.

82.2

82.1 Three-leaved false Solomon's seal: plant in flower. **82.2** Three-leaved false Solomon's seal: plant in fruit.

83

Starflower (*Lysimachia borealis*, formerly *Trientalis borealis*) is a perennial herb in the myrsine family, Myrsinaceae.

IDENTIFICATION This plant is 3–10 in. (7.5–25 cm) tall and has a whorl of several leaves at the top of the stem (83.1). The leaves are 1.5–4 in. (4–10 cm) long. From one to three fine stalks, 1–2 in. (2.5–5 cm) long, grow upward from the center of the whorl of leaves, each bearing a flower. The flowers typically have seven pointed and white petals and seven narrow and pointed green sepals (83.1, 83.2). The young, capsular fruit is green (83.2) and turns white (83.3).

OCCURRENCE · FAC This species occurs throughout our area in wooded fens, forested patches of bogs, and on hummocks at minerotrophic peripheral areas of bogs. It is especially widespread in non-wetland deciduous, mixed, and evergreen forests, and it ascends to subalpine peaty slopes.

83.1 Starflower: plant in flower. **83.2** Starflower: close-up of a flower, and of a young fruit surrounded by sepals. **83.3** Starflower: fruits.

84

Round-leaved sundew (*Drosera rotundifolia*) is one of five species of sundews in our area. It is a perennial herb in the sundew family, Droseraceae. Sundews are carnivorous, their prey largely consisting of small insects. They use droplets of mucilage produced by needlelike "tentacles" on the leaf blades to immobilize their prey. These slowly movable tentacles then press the prey against the leaf blade for digestion. Like pitcher plants, sundews obtain their energy by photosynthesis, and use carnivory to supplement the sparse supply of nutrients in the peat soil. See the Wikipedia *Drosera* page for more information.

IDENTIFICATION The leaves of round-leaved sundew radiate from the tip of a stem just under the surface (84.1), each leaf with a 0.6–2 in. (1.5–5 cm) petiole topped by a near-round blade that is 0.15–0.4 in. (0.4–1 cm) across, not counting the tentacles (84.1). The flower stem (scape) is 3–12 in. (7.5–30 cm) tall, generally less than 9 in. (23 cm) and is topped by flowers with white petals (84.2). The fruits (capsules) are light brown when mature (84.3).

OCCURRENCE · OBL This species occurs throughout our area on moist peat and organic soil at open and open-wooded sites at fens and bogs and at seeps and wet lake and stream shorelines.

84.2

84.3

84.1 Round-leaved sundew: two plants on sphagnum. **84.2** Round-leaved sundew: inflorescence with only top flower open. **84.3** Round-leaved sundew: fruits (capsules).

85

Spatulate-leaved sundew or intermediate sundew

(*Drosera intermedia*) is a carnivorous perennial herb in the sundew family, Droseraceae. See the round-leaved sundew page for further information on sundew carnivory.

IDENTIFICATION The leaves emerge from the subsurface stem in the form of a rosette (85.1), each petiole 1–2 in. (2.5–5 cm) long and topped by a spatulate blade 0.4–0.8 in. (1–2 cm) long (85.1, 85.2). The leaves open like a fern fiddlehead. The flower petals are white (85.1, 85.3). The fruits (capsules) are reddish brown when mature (85.4). Sundew species of our area with more elongate leaf blades include English sundew (*Drosera anglica*), slender-leaved sundew (*D. linearis*), and thread-leaved sundew (*D. filiformis*) (all OBL).

OCCURRENCE · OBL *Drosera intermedia* is found throughout our area at open and open-wooded sites on wet peat and in shallow pools at fens and bogs and along wet lake and stream shorelines. *Drosera anglica* and *D. linearis* are northern species of circumneutral fens, in our area only in Maine, Wisconsin, Michigan, Ontario, and Québec, and *D. anglica* also in New Brunswick. *Drosera filiformis* is more southern and east coastal, in our area represented only by small populations in Nova Scotia, Massachusetts, and New York.

85.2

85.3

85.1 Spatulate-leaved sundew: three plants, one of them with flower.
85.2 Spatulate-leaved sundew: parts of two plants showing some details
of the leaves. **85.3** Spatulate-leaved sundew: flowers. **85.4** Spatulate-leaved
sundew: fruits (capsules).

86

Swamp candles or swamp yellow loosestrife (*Lysimachia terrestris*) is a perennial herb in the myrsine family, Myrsinaceae.

IDENTIFICATION The erect stem is 10–40 in. (0.25–1 m) tall, often branched, and bears lanceolate, opposite leaves. The main leaves are 2–4 in. (5–10 cm) long, and shorter leaves occur toward the tips of the branches (86.1). This species propagates by sprouts from rhizomes, and by two other methods, seeds and bulblets. The five-petaled flowers are borne in a single elongate terminal cluster thought to resemble a candle (86.1). The petals are yellow, each one with two red dots at its base (86.2). The fruits (capsules) (86.3), green at first, dry out, turn brown, and split open to shed seeds. By late summer, some plants produce bulblets in the axils of the leaves lower on the stem (86.4). Bulblets may give rise to new plants, a type of vegetative reproduction (not from a flower).

OCCURRENCE · OBL This species is common and widespread throughout our area. It grows at open and open-wooded wetlands including fens, the peripheral minerotrophic parts of bogs, wet meadows, and lake, river, and stream shorelines.

86.1 Swamp candles: stem with inflorescence.
86.2 Swamp candles: flowers.
86.3 Swamp candles: fruits (capsules), most of them still green. Photograph: Glen H. Mittelhauser. **86.4** Swamp candles: bulblets. Photograph: Glen H. Mittelhauser.

87

Halberd-leaved tearthumb (*Persicaria arifolia*) is an annual herbaceous vine-like plant in the buckwheat family, Polygonaceae. It is one of the few annuals that occur in our peatlands.

IDENTIFICATION The young small plant is erect, but as it elongates to 3–6 ft. (0.9–1.8 m) it reclines on other plants. Reflexed prickles (87.1) help the plant to hold its aboveground position. The leaf blades are arrow shaped with divergent basal lobes. The flowers are green to white to pink, egg shaped, with petals that stay mostly closed, giving the flower a bud-like appearance (87.1). The fruit is a brown to black achene.

OCCURRENCE · OBL This species occurs throughout our area at the periphery of bogs and fens, in swamps and wet meadows, along lake and stream shorelines, and in ditches.

87.1 Halberd-leaved tearthumb: leaves, stems, and flowers.

88

Sweet white violet (*Viola blanda*) is a perennial herb in the violet family, Violaceae. Additional violet species rarely or sporadically occur in our peatlands, but this is the species most likely to be encountered, in my experience.

IDENTIFICATION The plant spreads by slender stolons or runners, forming colonies. The leaves are broadly heart shaped (88.1). The fragrant white flowers are borne on long, nodding peduncles (88.1) and are 0.3–0.5 in. (0.8–1.3 cm) long. The lowest three petals have purple veins near their base, and the upper two petals are bent back to varying degrees (88.1, 88.2). At first, the ovoid fruits (capsules) are purple with lighter flecks (88.3). The fruit turns tan as it dries, splits open (88.4), and sheds tiny seeds.

OCCURRENCE · FACW This species occurs throughout our area, usually at shaded sites on hummocks at the periphery of fens and bogs; also in swamps, at lake and stream shorelines, moist woods, moist roadsides, and moist mowed fields.

88.1 Sweet white violet: plants with flowers on nodding peduncles. 88.2 Sweet white violet: close-up of flower. The purple veins on the lower lateral petals are hard to see. 88.3 Sweet white violet: fruit (capsule) on nodding peduncle. 88.4 Sweet white violet: four valves of an open capsule after seeds have been shed; one valve (on right) is reflexed and almost entirely hidden by another.

89

Northern water-horehound (*Lycopus uniflorus*) is a perennial herb in the mint family, Lamiaceae.

IDENTIFICATION The plant is 8–30 in. (20–76 cm) tall and has a square stem. The leaves (89.1) are opposite, elongate (to about 4 in. or 10 cm), gradually narrowed at both ends, and coarsely toothed except for the basal third (approx.). Compact clusters of white flowers occur in the leaf axils (89.1). The flower petals are fused at the base, forming a bell-like shape with four flaring lobes at the tip. The upper lobe is more or less incised and appears as two lobes in many specimens—forming an ostensible five lobes. This species may be difficult to distinguish from Virginia water-horehound (*L. virginicus*—OBL). Typically, *L. virginicus* has a more abruptly narrowed basal part of the leaf, virtually forming a petiole, and often the leaves are purple hued, especially on the underside. There are four petal lobes, and these lobes generally are less flared than in *L. uniflorus*, or not flared at all. *Lycopus virginicus* typically does not produce tubers, whereas *L. uniflorus* consistently does so. The fruits of both species are nutlets in groups of four.

OCCURRENCE · OBL *Lycopus uniflorus* is widespread in North America and occurs throughout our area at wet open and open-wooded sites at the periphery of fens and bogs. It also occurs widely on wet mineral soils in swamps and meadows, at seeps, and along lake, river, and stream shorelines. *Lycopus virginicus* occurs in similar habitats in the states and provinces covered here but is absent from New Brunswick and Nova Scotia.

89.1 Northern water-horehound: plants in flower.

90

Tall white-aster or flat-topped white aster (*Doellingeria umbellata* var. *umbellata*) is a perennial herb in the aster family, Asteraceae.

IDENTIFICATION This plant grows to 6.5 ft. (2 m) tall, is branched (some shorter plants may be unbranched), and has a reddish to brownish main stem. The leaves are alternate and are 1.5–6.5 in. (4–17 cm) long (90.1). The flower heads number from 20 to 200 per plant, fewer in the north, and are arranged in more or less flat clusters (90.1, 90.2). Each flower head has white ray flowers, from 6 to 14 per head, with blades 0.25–0.4 in. (0.6–1.0 cm) long. The disk flowers are yellow green, many per disk, with club-shaped projections (90.2). The small dry fruits (achenes) have many bristles that facilitate wind dispersal (90.3).

OCCURRENCE · FACW This species occurs throughout our area at the peripheral parts of fens and bogs and in marshes, wet to moist meadows and fields, and at forest edges and roadsides.

90.1 Tall white-aster: top part of a single plant in flower. **90.2** Tall white-aster: close-up of flowers. **90.3** Tall white-aster: a head in fruit, showing achenes with bristles. Photograph: Marilee Lovit.

91

Bog willow-herb or American marsh willow-herb (*Epilobium leptophyllum*) is a perennial herb in the evening primrose family, Onagraceae.

IDENTIFICATION The plant is 12–24 in. (30–60 cm) tall and has narrow revolute leaves. Small leaves arise in the axils of many of the larger ones (91.1). The surfaces of the young leaves have fine, short, pressed-down hairs. The plants at shaded sites (91.1, 91.2) may be less robust than those at open sites. The four white or pink petals of the flower emerge from the tip of a tubular structure (91.1). The elongate fruits (capsules) split open lengthwise to release numerous small seeds, each seed bearing long white to tan hairs for wind dispersal (91.2). The similar species, marsh willow-herb (*E. palustre*—OBL) has wider leaves essentially lacking hairs on the top surface.

OCCURRENCE · OBL *Epilobium leptophyllum* occurs at the periphery of fens and bogs, and at marshes, wet meadows, lake and stream shorelines, and wet ditches throughout our area. *Epilobium palustre* occupies similar habitats but is absent from Ohio, Indiana, and Illinois.

91.2

91.1 Bog willow-herb: plant with tubular floral structures showing emergent petals in two cases. Shaded site. **91.2** Bog willow-herb: linear fruits (capsules) partly open and shedding hairy seeds. Shaded site.

92

Rough wood-aster or low rough aster (*Eurybia radula*) is a perennial herb in the aster family, Asteraceae.

IDENTIFICATION The single stem grows to 45 in. (1.1 m) tall but only to 32 in. (0.8 m) in our peatlands. The alternate, veiny, elongate, pointed, and more or less serrate leaves lack petioles and are up to about 4 in. (10 cm) long (92.1, 92.2). The plant has 1 to 40 flower heads clustered on top (92.1, 92.2). The heads have pale to medium blue-purple rays and many tiny, yellow disk flowers (92.1, 92.2). The fruit is an achene with bristles for wind dispersal.

OCCURRENCE · OBL This species occurs at open and open-wooded sites and edges in fens and the minerotrophic periphery of bogs; also at lake, river, and stream shorelines and other wet to moist locations. It is absent from Wisconsin, Michigan, Illinois, Indiana, and Ohio but occurs in all the other states and provinces of our area. It is imperiled or has been extirpated over much of the southeastern part of our area and is listed as endangered in Connecticut, New Jersey, and New York.

92.2

92.1 Rough wood-aster: top of a plant in flower. **92.2** Rough wood-aster: top of a plant with more numerous flowers, viewed from above.

92.1

93

Tufted yellow-loosestrife (*Lysimachia thyrsiflora*) is a perennial herb recently placed in the myrsine family, Myrsinaceae (Haines 2011), but formerly in the primrose family, Primulaceae.

IDENTIFICATION The erect stem of the plant is 8–32 in. (0.2–0.8 m) tall and has opposite to sub-opposite or whorled, lanceolate leaves 2–5 in. (5–13 cm) long (93.1). The leaves are dotted (93.2). Ovoid inflorescences, 0.4–1.2 in. (1–3 cm) long with yellow flowers, are borne at the ends of 0.8–1.6 in. (2–4 cm) stalks (peduncles) that arise from the axils of leaves at the middle part of the stem (93.1, 93.2). The fruits (capsules) occur in ovoid bunches corresponding to the inflorescences, each fruit 0.08–0.12 in. (2–3 mm) long (93.3) and becoming brown as it dries out before splitting open to shed its seeds. When not in flower, tufted yellow-loosestrife might be confused with purple loosestrife (*Lythrum salicaria*—OBL), an exotic invasive plant of wetlands in our area, but tufted loosestrife has a round stem, and purple loosestrife a square stem.

OCCURRENCE · OBL This species occurs throughout our area at open and wooded fens, the minerotrophic periphery of bogs, and marshes, lakeshores, and shallow water.

93.1 Tufted yellow-loosestrife: plants with inflorescences. **93.2** Tufted yellow-loose-strife: closer view of inflorescences. **93.3** Tufted yellow-loosestrife: fruits. Photograph: Keir Morse.

93.2

Ferns

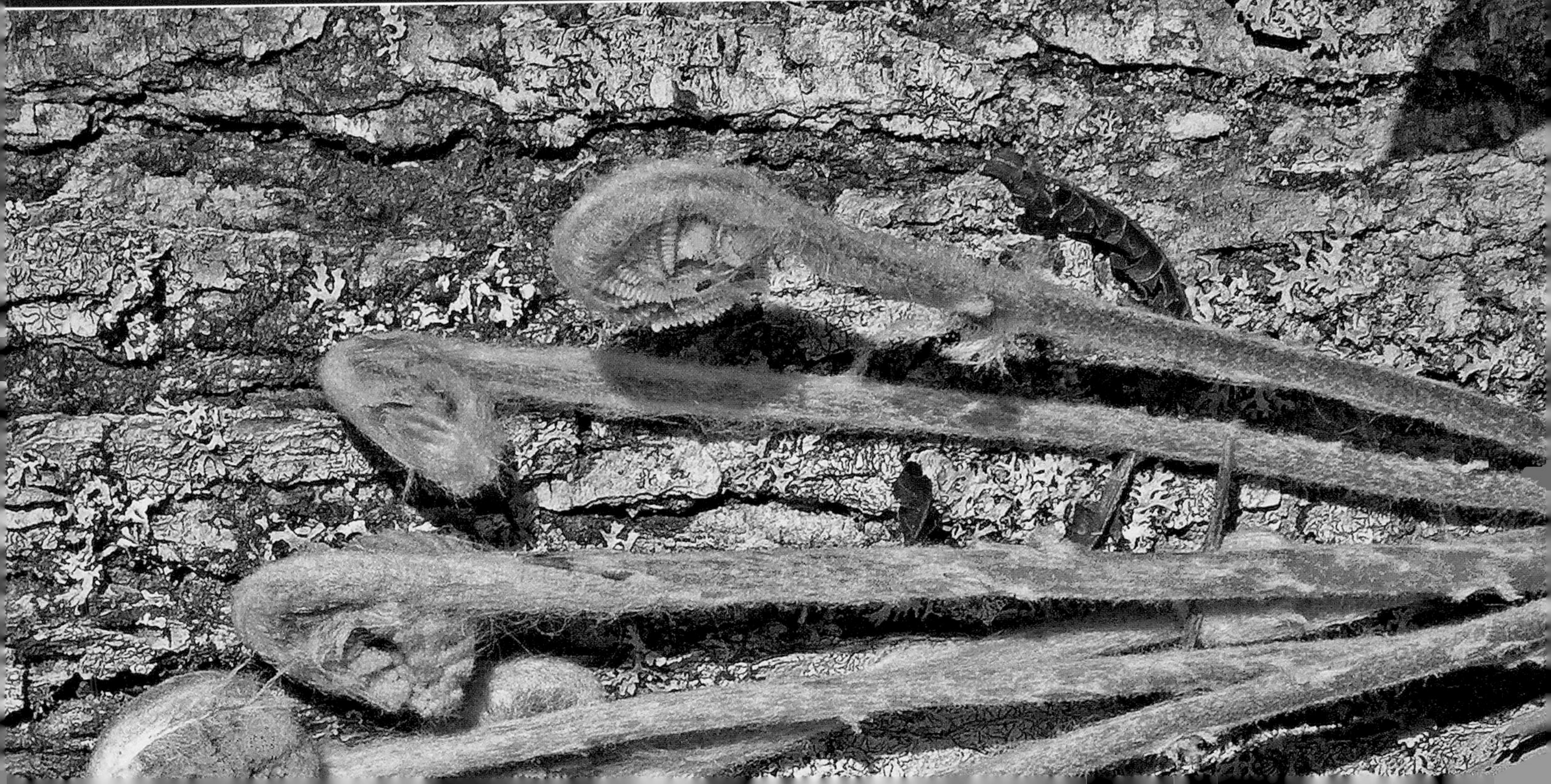

Cinnamon fern (*Osmundastrum cinnamomeum,* formerly *Osmunda cinnamomea*) is a perennial in the royal fern family, Osmundaceae. This and other ferns are spore-producing vascular plants lacking flowers, fruits, and seeds.

IDENTIFICATION In spring, groups of fronds arise from over-wintering underground or surface stems (rhizomes). At first, the fronds are furled as densely fuzzy fiddleheads (94.1, 94.2). These unfurl and expand into two types of fronds, fertile and infertile. Fertile fronds are highly specialized for spore production (94.3). They expand most rapidly, become 1.5–3 ft. (0.46–0.9 m) tall, darken to a cinnamon color, and release their spores. Soon after releasing spores they start wilting down to the ground. The infertile fronds (94.3) grow taller (to 5 ft.; 1.5 m) than the fertile fronds, lose their fuzz except for tufts at the bases of frond subdivisions (pinnae), and remain standing for the rest of the season.

OCCURRENCE · FACW Cinnamon fern occurs throughout our area at open and wooded sites in fens and the minerotrophic periphery of bogs, various mineral-soil wetlands, moist woods, and lake and stream shorelines.

94.1 Cinnamon fern: fiddleheads in three groups. These are partially surrounded by young sensitive fern fronds and skunk cabbages. **94.2** Cinnamon fern: close-up of fiddleheads adjacent to a red maple tree. **94.3** Cinnamon fern: upper parts of two infertile fronds not quite fully expanded, and three fertile fronds. This is the same plant as in photo 2, but later in spring.

95

Marsh fern (*Thelypteris palustris* var. *pubescens*), in the The-
lypteridaceae family, is a perennial that overwinters as rhizomes.
The fronds are sensitive to first frost.

IDENTIFICATION The fronds arise from black rhizomes, are up to
28 in. (0.7 m) tall, but generally only 14–24 in. (0.35–0.6 m) tall
in our peatlands. The blade is widest about two-fifths up from the
bottom (95.1). The basal stalk (stipe) is blackish, and the midrib
of the blade is medium to dark gray or gray green and typically
minutely hairy. Fertile fronds are similar in form to infertile ones
but may be taller. Their edges are revolute (95.2, 95.3), giving frond
subdivisions a slenderer outline than those of the infertile frond.
The underside develops many small structures called sori (95.2,
95.3), each containing the even smaller spore-producing units (sporangia). Each sorus is covered by an um-
brella-like structure called an indusium. The sori are
light green at first (95.2, 95.3), then darken and lose
their indusia, becoming brown black when ready to
release their spores.

OCCURRENCE · FACW This species occurs throughout our area at open and wooded sites in fens and
the minerotrophic periph-
ery of bogs, in marshes, swamps, moist meadows
and ditches, and along lake and stream shorelines.

95.1 Marsh fern: infertile frond. **95.2** Marsh fern: part of a fertile frond. Photograph: Glen H. Mittelhauser. **95.3** Marsh fern: smaller part of a fertile frond showing sori. Photograph: Robert W. Smith.

95.1

96

Royal fern (*Osmunda regalis* var. *spectabilis*) is a perennial in the royal fern family, Osmundaceae.

IDENTIFICATION Royal fern is the tallest (to 6 ft.; 1.8 m) fern species in our area. The fronds are twice-divided (96.1). In spring, the fronds arise from near-surface rhizomes as reddish-tan fiddleheads (96.2). The fiddleheads bear a cottony fuzz at first but quickly lose it and their reddish-tan color as they unfurl and expand. Fertile, spore-producing parts are formed at the tips of the fronds, green at first (96.1), then turning brown as they mature and shed spores (96.3). In the fall, the fronds die down, leaving the overwintering rhizomes.

OCCURRENCE · OBL Royal fern occurs throughout our area at wooded to open sites in fens and the minerotrophic periphery of bogs. It also occurs in swamps and along lake and stream shorelines.

96.1 Royal fern: fronds, some (center and left) bearing green (immature) spore-producing parts at their tops. **96.2** Royal fern: fiddleheads. **96.3** Royal fern: fertile part of frond.

97

Sensitive fern (*Onoclea sensibilis*), in the Onocleaceae family, is named for the sensitivity of its sterile fronds to frost. It is also named bead fern because of the appearance of the fertile fronds. It is a perennial with overwintering rhizomes.

IDENTIFICATION This fern has two types of fronds, infertile and fertile. Infertile fronds grow to 40 in. (1 m) long but generally are less than 30 in. (0.76 m) long in our peatlands. The central vein of the frond is winged (97.1). The typically shorter fertile fronds are highly specialized, with bead-like structures containing sporangia. These fertile fronds are unusual for ferns in our region in remaining upright through the winter. The beads are green at first (97.2), turn black (97.3), and overwinter before splitting open to release spores in the spring.

OCCURRENCE · FACW Sensitive fern occurs throughout our area at open and wooded sites in fens and the minerotrophic periphery of bogs. It also occurs in a wide range of mineral-soil wetlands, ditches, moist meadows and fields, and at lake, river, and stream shorelines.

97.1 Sensitive fern: infertile frond. **97.2** Sensitive fern: infertile and immature fertile frond. **97.3** Sensitive fern: top of near-mature fertile frond showing "beads" almost ready to overwinter.

98

Woodland horsetail or wood horsetail (*Equisetum sylvaticum*) is a perennial plant in the horsetail family, Equisetaceae. Like the ferns, it is a spore-producing vascular plant lacking flowers, fruits, and seeds.

IDENTIFICATION Woodland horsetail spreads by underground stems (rhizomes) that give rise to upright stems 6–24 in. (15–61 cm) tall aboveground (98.1). These stems are of two types, infertile (98.2) and fertile (98.3). Fertile stems appear first in the spring and bear a spore-producing cone-like structure (strobilus) on top (98.3). As green branches emerge in whorls at the nodes of these stems, and elongate, the strobilus matures, sheds its spores, and falls off. Infertile stems also bear whorls of branches at the nodes (98.2), each branch fine and subdivided. Both types of stems die down to the ground in late summer, and the plant overwinters as rhizomes. Another horsetail species that may be encountered in our peatlands is river horsetail, *E. fluviatile* (OBL). It has one spikelike stem with relatively short and typically undivided side branches, and this stem may grow as tall as 40 in. (1 m).

98.2

98.3

98.1 Woodland horsetail: group of plants in summer. The blunt-topped one in center originated as a fertile plant that in spring bore a spore-producing cone on top. After shedding its spores, the cone fell off and the branches elongated. See text and photo 3. **98.2** Woodland horsetail: infertile stem with whorled branches. **98.3** Woodland horsetail: fertile stem with mature strobilus and developing branches.

OCCURRENCE · FACW Woodland horsetail occurs throughout our area except Indiana. It grows at relatively dry-surfaced, wooded parts of fens and minerotrophic parts of bogs, including on hummocks at the peatland periphery. It occurs more widely at moist upland forests and edges. River horsetail also occurs throughout our area but at much wetter sites and typically in standing or slow-flowing water, as at the edges of fens, in marshes, streams, and the shallows of ponds and lakes.

Additional Peatland Species: An Annotated List

If you are at a bog or fen in our region and are confident that the plant you are looking at is not one of the 98 featured or 34 comparative species in the descriptions, take a photo of it if you can, or at least make note of the plant's characteristics. Check for possible matches from the following annotated list of 23 additional peatland species. Consult the following websites for photos and descriptive information that will help you pin down the identification: https://gobotany.newenglandwild.org; http://michiganflora.net; and https://plants.usda.gov/java/.

Inclusion of a species in this section is based on its designation as diagnostic or characteristic of one or more peatland plant communities, peatland natural communities, or peatland ecological communities (terminology dependent on political jurisdiction) at state and province natural resource agency and ministry websites. Although these species characterize one or more peatland natural communities, they are not as common or widespread in the peatlands of our area as the large majority of featured species.

Trees

BLACK TUPELO (*Nyssa sylvatica*—FAC) is a deciduous tree with alternate, simple, and smooth-edged leaves, and a many-branched stout trunk. It is a southern species that is absent from the northern part of our area. In our area it mostly grows in wooded fens and swamps, but farther south it also grows widely at moist uplands.

EASTERN WHITE PINE (*Pinus strobus*—FACU) is an easily recognizable, temperate evergreen tree of our entire area and beyond, with needles in bunches of five. It occurs, typically sparingly, in many of our peatlands.

JACK PINE (*Pinus banksiana*—FACU) is a straggly tree with short needles in bunches of two. It occurs in all our jurisdictions and northward. In our area it grows mostly in dry woodlands that originated from disturbance, commonly by fire, but also ventures into open fens and bogs.

PITCH PINE (*Pinus rigida*—FACU) is an evergreen tree with needles in bunches of three, and frequently has a crooked trunk. It grows in dry woodlands, often of fire origin, and also in swamps and at the edges of fens and bogs.

YELLOW BIRCH (*Betula alleghaniensis*—FAC) is a deciduous tree with alternate, simple leaves with toothed edges. It has yellowish peeling bark (except very old trunks) and sap with a wintergreen aroma. It occurs in all our jurisdictions in swamps and wooded fens, and commonly in non-wetland forests.

Shrubs

COASTAL SWEET-PEPPERBUSH (*Clethra alnifolia*—FAC) is a tall deciduous shrub with simple, alternate leaves with toothed edges. It grows in the understory of swamps and wooded fens and other habitats of our Atlantic Seaboard states. In the United States it is absent west of Pennsylvania. In Canada it occurs only in Nova Scotia.

SPICY WINTERGREEN, EASTERN (*Gaultheria procumbens*—FACU), is an aromatic evergreen trailing shrub with partly herbaceous stems, and leaves that are alternate, simple, shiny, and leathery. It occurs in all our jurisdictions at relatively dry microhabitats in our swamps, wooded fens and bogs, and more widely at upland habitats.

Herbs

BEAKED SPIKESEDGE (*Eleocharis rostellata*—OBL) is a sedge with a single terminal spikelet and leafless stem. It occurs in all our jurisdictions except Vermont, New Hampshire, New Brunswick, Prince Edward Island, and Québec. In the eastern part of our area it grows in tidal saltmarshes; in the western part of our area in circumneutral to alkaline fens and at calcareous shorelines.

FEW-NERVED COTTONSEDGE (*Eriophorum tenellum*—OBL) is a sedge with a single short bract subtending an inflorescence with drooping cottony spikelets. It is a northern species that reaches its southern limit in our area. It grows in open bogs and fens and in conifer-wooded fens and swamps. It occurs in all our jurisdictions except Indiana and Ohio.

INLAND SEDGE (*Carex interior*—OBL) grows in tufts and has perigynia with a convex upper half. It occurs in all our jurisdictions at open and wooded circumneutral and alkaline fens, meadows, conifer (often cedar) swamps, and calcareous shorelines and seeps.

LAKESIDE SEDGE (*Carex lacustris*—OBL) is a fairly tall sedge with reddish-brown basal leaf sheaths, leaves with an M-shaped cross section, and cylindrical spikes of ascending perigynia. It occurs in all our jurisdictions at wet fen and bog margins, marshes, shores, and floodplains of rivers and lakes, and open swamps.

SOFT-LEAVED SEDGE (*Carex disperma*—OBL) is a small sedge with soft and weak leaves. It occurs in all our jurisdictions in wooded fens and swamps and wooded bog margins. Although primarily a plant of the shade, it also occurs in wet meadows.

SWOLLEN-BEAKED SEDGE (*Carex utriculata*—OBL) has cylindrical spikes with spreading, inflated perigynia. It is a widespread species in all our jurisdictions. It grows in colonies at open fens, bog margins, sedge meadows and marshes, wet thickets and open swamps, and along lake, stream, and river shores.

THREE-WAY SEDGE (*Dulichium arundinaceum*—OBL) is distinctive in the emergence of its leaves in three near-perfect lines along a nearly round stem, as can be clearly seen from the top of the plant. It occurs in all our jurisdictions at marshy stream, river, and lake shores, marshes, and wet areas of open fens and bog margins, including in shallow water.

TUSSOCK SEDGE (*Carex stricta*—OBL) can form large, tall tussocks. Its basal leaf sheaths shred into cross fibers that look like a ladder. It occurs in all our jurisdictions in open wet areas of fens, bog margins, marshes, and wet meadows, seasonally flooded sites like marshy stream and lake shores, and often forms extensive stands.

YELLOW-GREEN SEDGE (*Carex flava*—OBL) is a medium-size sedge with ovoid pistillate spikes with yellow to yellow-green and bent-down perigynia. It reaches its southern limit in our area (also in Virginia) but is absent from Indiana. It grows in open circumneutral and alkaline fens, marshes, and meadows and open-wooded fens and swamps, often marly.

CANADA RUSH (*Juncus canadensis*—OBL) occurs in tufts, has leaves that are round in cross section, and stiff upright stems that bear spheroid clusters of flowers. It occurs in all our jurisdictions at acidic to alkaline, open and open-wooded fens and bog margins, marshes, ditches, shorelines, and slightly brackish upper borders of coastal tidal marshes.

SPIKE MUHLY (*Muhlenbergia glomerata*—OBL) is a grass with a slender, compact inflorescence, is often purple tinged, and spreads by scaly rhizomes. It occurs in all our jurisdictions in fens, marshes, meadows, moist fields, open swamps, and calcareous shores.

AMERICAN TWINFLOWER (*Linnaea borealis* ssp. *americana*—FAC) is an herb with opposite leaves on trailing stems and has upright flower stalks with pairs of nodding, typically pink flowers. It is at its southern lowland limit in our southern tier of states. It grows in wooded fens and bogs at relatively dry microsites and in non-wetland forests and borders.

BOG NODDING-ASTER (*Oclemena nemoralis*—OBL) is a slender herb with narrow leaves that are rough above and pubescent below. It bears one to a few heads at the top, each with 13 to 27 pink to purple rays. It is present in all our jurisdictions except Ohio, Indiana, Illinois, and Wisconsin at acidic to alkaline fens, bogs, meadows, shorelines, and moist rock crevices.

MARSH CINQUEFOIL (*Comarum palustre*—OBL) (formerly *Potentilla palustris*), with trailing lower stem, then ascending, has five to seven toothed leaflets per compound leaf, and showy red-purple flowers. It occurs in all our jurisdictions in fens and fen lake borders, wet margins of bogs, marshes, and meadows, and along stream and river shores.

Ferns

CRESTED WOOD FERN (*Dryopteris cristata*—OBL), with slender fronds narrowing toward the base, has leaflets angled parallel to the ground, and round to kidney-shaped sori on the fertile fronds. It is present in all our jurisdictions in fens and bogs on hummocks, in swamps and wet thickets, wetland margins, sedge meadows, stream and lake shores, and seeps.

VIRGINIA CHAIN FERN (*Woodwardia virginica*—OBL) is a tall (to 4.6 ft.; 1.4 m) fern that grows in lines from creeping rhizomes. Sori are elongate, with a set along and parallel to the leaflet's central vein. It is present in all our jurisdictions except Wisconsin in acidic wetlands including fens, bogs, marshes, swamps, lake shores, and ditches.

In this section I list peatlands with boardwalks in 18 of our 19 states and provinces (map 1). I found no boardwalks in the peatlands of Rhode Island. Boardwalks in peatlands and other wetlands serve three purposes: (1) they provide ease of access to wet ecosystems that otherwise would be very difficult to traverse, (2) they protect the wetland vegetation and upper peat by confining visitors to a single route that eliminates trampling, and (3) they provide a pleasurable and educational experience, thereby helping to build a public constituency for the protection and conservation of these important ecosystems.

Peatlands are very sensitive to trampling and may take decades or longer to recover once the soft upper peat is damaged. Although boardwalks do some harm by shading the plants, this impact is confined to a narrow strip. If improperly designed, however, boardwalks may damage the upper peat and may impede the horizontal flow of surface and near-surface water. But, on balance, boardwalks are a necessity if large numbers of outdoor enthusiasts, ecotourists, and nature lovers are to visit peatlands without inflicting even more serious damage.

The abundance of peatland boardwalks varies greatly from state to state and province to province, depending not only on the wide differences in peatland abundance, but also on differences between programs and priorities of government conservation and land-management agencies and ministries, and of nongovernmental organizations (NGOs) concerned with the environment. For example, although the state of Maine has many more peatlands, and a much larger total area of peatland, than the adjacent and much smaller state of New Hampshire, New Hampshire has many more peatland boardwalks.

Boardwalks in our states and provinces have been built by, and are being managed by, a wide range of entities, including federal (U.S.), national (Canada), state, and provincial agencies and ministries, town governments, large NGOs with state chapters like the Nature Conservancy in the United States, and by local NGOs. Peatland boardwalks are located in national, provincial, and state parks, forest and wildlife reserves, town parks and reserves, NGO parks and reserves, and on private lands. In all cases, peatlands containing boardwalks are in some form of protection as natural area reserves.

I gathered information on peatlands with boardwalks, and on the boardwalks themselves, by questionnaires mailed to government agencies and ministries, searches of government and NGO websites, and by telephone and e-mail communications with these agencies, ministries, and NGOs. Although I tried to locate all peatlands with boardwalks in each state and province, I make no claim to success in that endeavor. Additionally, available information on the peatlands and their boardwalks is incomplete for some sites.

When using the following list, keep in mind that a peatland's name, such as Severin's Bog, is not a reliable indicator of whether it is a true bog or a fen, nor is a name like Lady Swamp a reliable indicator of whether it is a mineral-soil wetland or a wooded fen. Scientific designation as bog or fen, swamp or wooded fen, or more specific designation is not given in the information available for most sites. But for many peatlands I was able to add these designations from characteristics revealed by satellite imagery on https://www.google.com/maps/ and from plant communities described on nature reserve websites. The importance of such designations for this book is that, together with geographic location, they prepare visitors on plant species to look for. In similar fashion, knowledge of habitat and geographic area prepares birders on bird species to look for. Additionally, as for birders and their checklists, expectations are directly facilitated by plant species lists for peatlands, at least for those species that can be seen from the boardwalk. However, with notable exceptions such lists have been unavailable.

The following list of peatland boardwalks is organized alphabetically by state and province, and by site within each jurisdiction. Each listing briefly describes the peatland and the boardwalk, including boardwalk structural type. Structural types fall into three broad categories: (1) a structure with decking of parallel boards attached to a supporting frame consisting of joists, the frame raised on footings or floats above the peatland surface; (2) as number 1, but raised above the surface on pilings, suitable only where the peat is shallow; and (3) the simplest and least costly, long planks or split logs in line with boardwalk direction, typically two planks or logs side by side and attached to short supporting crosspieces on the peatland surface. Site descriptions also give ownership and/or management agency, ministry, or NGO, and Internet sources (all accessed March–July 2015) with additional information including travel directions. Finally, I indicate if the boardwalk, along with any access trail, is wheelchair accessible.

Each of the sites is numbered, and these numbers appear on map 1, enabling the reader to see the approximate location of each site within the states and provinces.

Connecticut

1. Black Spruce Bog Natural Area Preserve, Mohawk Mountain State Park, Cornwall and Goshen. A boardwalk about 500 ft. (150 m) long, consisting of two planks slightly elevated above the surface, traverses this largely coniferous wooded fen. Information kiosk. Owned and managed by the Connecticut Department of Energy and Environmental Protection, State Parks and Forests.
   http://www.ct.gov/deep/cwp/view.asp?A=2716&Q=325060
   https://suite.io/karen-bartomioli/5qf72n9

Illinois (Glaciated Part Only)

2. Volo Bog State Natural Area, Ingleside, has a 47.5-acre "quaking bog" that is largely an unwooded fen with an open-water center in a kettle. An approximately 0.4 mi. (0.7 km) trail has a decked and raised boardwalk loop for about half its length. Visitor center and interpretive program. Owned and managed by the Illinois Department of Natural Resources.
   http://dnr.state.il.us/lands/landmgt/PARKS/R2/Volobog.htm

Indiana (Glaciated Part Only)

3. Pinhook Bog, in Indiana Dunes National Seashore, Michigan City, is accessed by a 0.32-mi. (0.5 km) trail leading to a 0.13-mi. (0.2 km) boardwalk. The peatland occupies a kettle. Guided walks for school groups are available by advanced arrangement. Owned and managed by the National Park Service.

> http://www.nps.gov/indu/planyourvisit/pinhook-bog.htm
> http://www.nps.gov/indu/learn/education/upload/PB_il.pdf

Maine

4. Hidden Valley Nature Center, Jefferson, has a peatland in a kettle that is accessed by a 220-ft. (67 m) decked and raised boardwalk. A viewing platform has interpretive information. Guided walks. Owned and managed by the Hidden Valley Nature Center.

> http://hvnc.org/2012/12/27/hidden-valley-in-jefferson-for-all
> -to-see/

5. Orono Bog Boardwalk, Orono and Bangor, is a 0.8-mi. (1.3 km) decked and raised boardwalk in a raised bog. Interpretive signs and guided walks. Checklists of plants and birds are available at the website. This boardwalk (shown in map 2) is pictured on the book cover and at the beginning of this chapter. Owned by the University of Maine and the City of Bangor and managed in collaboration with the Orono Land Trust. Wheelchair accessible.

> http://umaine.edu/oronobogwalk/

6. Quoddy Head State Park, Lubec, has a coastal bog accessed by a 0.6-mi. (1 km) trail. The trail ends in the bog, with a 0.13-mi. (0.2 km) decked boardwalk loop with interpretive signs. This bog's flora includes baked-apple berry and black crowberry. Owned and managed by the Maine Bureau of Parks and Lands.

> http://www.mainetrailfinder.com/trails/trail/quoddy-head
> -state-park

7. Saco Heath, Saco, may be the southernmost raised bog in eastern North America. It has a 0.5-mi. (0.8 km) nature trail with a decked and raised boardwalk. Owned and managed by the Nature

Conservancy. Information including a brochure for the self-guided nature trail at the website:

http://www.nature.org/ourinitiatives/regions/northamerica
/unitedstates/maine/placesweprotect/saco-heath-preserve
.xml

8. Salmon Brook Lake Fen in a reserve managed by the Maine Bureau of Parks and Lands in Perham is a ~220 acre (~90 ha) wooded and open fen with exceptional plant diversity including six rare species (maine.gov link). It has a ~125 ft (~38 m) boardwalk with viewing platform plus several shorter stretches along trails. Access is described in the bangordailynews.com link.

https://www.maine.gov/dacf/mnap/reservesys/salmonbrook.
htm

https://actoutwithaislinn.bangordailynews.com/2016/08/02/
one-minute-hikes/1-minute-hike-salmon-brook-lake-trails-
in-perham/

9. Ferry Beach State Park, Saco, features the Tupelo Trail. It includes a 0.4-mi. (0.65 km) raised, decked boardwalk over the wettest part of a black tupelo and red maple wooded fen. This fen is near the northeastern limit of black tupelo. Owned and managed by the Maine Bureau of Parks and Lands.

www.maine.gov/ferrybeach

http://www.maine.gov/dacf/mnap/assistance/hikes/ferry
_beach.pdf

http://www.mainetrailfinder.com/trails/trail/ferry-beach
-state-park/

Massachusetts

10. Atlantic White Cedar Swamp Trail, Wellfleet, Cape Cod National Seashore, is a 1.25-mi. (2 km) trail that includes a 0.4-mi. (0.65 km) decked and raised boardwalk in a wooded fen in a kettle with Atlantic white cedar, red maple, and coastal sweet-pepperbush. Owned and managed by the National Park Service. Wheelchair accessible.

http://www.nps.gov/caco/planyourvisit/upload/Atlantic

WhiteCedarcolor.pdf

11. Hawley Bog Preserve, Hawley, has a kettle peatland accessed by a 0.13-mi. (0.2 km) boardwalk consisting of two slightly elevated planks. Owned by the Nature Conservancy (TNC) and the Five Colleges Consortium, and managed by TNC.

> http://www.nature.org/ourinitiatives/regions/northamerica
> /unitedstates/massachusetts/placesweprotect/hawley-bog
> -preserve.xml

12. Pine Hole Bog Nature Trail, Ward Reservation, Andover and North Andover, is a self-guided nature trail with a 0.13-mi. (0.2 km) boardwalk in a kettle peatland. The numbered interpretive stations are accompanied by a booklet available at the second website. Owned by Ward Reservation and managed by the Trustees for Reservations.

> http://www.thetrustees.org/places-to-visit/northeast-ma
> /ward-reservation.html#t1
> http://www.thetrustees.org/assets/documents/places-to-visit
> /trailmaps/PineHoleBog.pdf

Michigan

13. Bishop's Bog Preserve Trail is part of a 6-mi. (10 km) trail system at Portage South-Central Greenway, with about 1.5 mi. (2.4 km) of decked floating boardwalks in fens and marshes. The Bishop Bog part is about 0.9 mi. (1.5 km) long. Owned and managed by the City of Portage.

> http://www.portagemi.gov/Departments/PRSCS/Parks
> Recreation/ParksAmenitiesListing/BishopsBogPreserve
> .aspx
> http://www.michigantrailmaps.com/member-profile/3/218/

14. Camp Newaygo Wetland Trail, Newaygo, is a 0.5-mi. (0.8 km) two-planked boardwalk on the peatland surface. First it traverses a wooded fen, then an open fen. It is owned and managed by Camp Newaygo.

> http://www.campnewaygo.org/WetlandTrail.aspx
> http://outdoormichigan.org/feature/3104

15. Mud Lake Bog Nature Preserve, Buchanan, has a 0.4-mi. (0.6 km) trail, most of it a boardwalk consisting of three planks on the surface. This walkway traverses wooded and open parts of a kettle peatland. Owned by Buchanan Township and managed by the nearby Fernwood Botanical Garden and Nature Preserve and the Love Creek County Park and Nature Center.

> http://outdoormichigan.org/feature/4553
> http://www.trails.com/tcatalog_trail.aspx?trailid
>   =HGM056–034

16. Presque Isle Park Bog Walk, Marquette, features an approximately 0.5-mi. (0.8 km) trail, most of which is a boardwalk. Interpretive signs. Owned by the City of Marquette and managed by the city and the Moosewood Nature Center.

> http://www.waymarking.com/waymarks/WMCC3M_Presque
>   _Isle_Park_Bog_Walk_Mar
> http://www.mqtcty.org/Maps/PresqueIsle.pdf

17. Raspberry Island Trail, Isle Royale National Park, is an approximately 2-mi. (3.2 km) outing, including a short canoe or boat ride from Isle Royale. The trail on the island traverses uplands and peatlands, including a slightly raised two-plank boardwalk in an acidic fen. Self-guided with interpretive signs. Owned and managed by the National Park Service.

> http://www.nps.gov/isro/planyourvisit/outdooractivities.htm

18. West Lake Wetland Walk, in West Lake Nature Preserve at Portage, features an approximately 1-mi. (1.6 km) trail through a variety of wetlands, including a "bog," in this case an acidic fen. Stretches of floating boardwalk may make for wet walking at times. Owned and managed by the City of Portage. A self-guided nature trail brochure may be downloaded from the website:

> http://www.portagemi.gov/Departments/PRSCS/Parks
>   Recreation/ParksAmenitiesListing/WestLakeNature
>   Preserve.aspx

New Brunswick

19. Caribou Plain Bog, in Fundy National Park in Alma, is an

acidic peatland with an approximately 0.5-mi. (0.8 km) boardwalk of raised decking. The boardwalk is part of the 1.3-mi. (2.1
km) Caribou Plain Trail. Owned and managed by Parks Canada.
Wheelchair accessible.

> https://www.youtube.com/watch?v=PCylcVrdNwg

20. Eagle Hill Bog, in Roosevelt Campobello International Park
(RCIP), Welshpool, has a boardwalk consisting of decking units
placed on the bog surface. It is a 0.6-mi. (1 km) round-trip walk.
Coastal bog species like baked-apple berry and black crowberry
are present. Interpretive signs. Owned and managed by RCIP. The
second website is a guide booklet.

> http://www.mainetrailfinder.com/trails/trail/roosevelt
> -campobello-international-park-eagle-hill-bog
> http://www.fdr.net/wp-content/themes/campobello/images
> /the-bogs-information-guide.pdf

21. Kelly's Bog, Kouchibouguac National Park, Kent County, has
a Bog Trail that is about 0.6 mi. (1 km) one way. It includes approximately 0.45 mi. (0.75 km) of decked boardwalk on the surface of
a pristine bog. There are interpretive panels and an observation
tower. Owned and managed by Parks Canada.

> http://www.pc.gc.ca/eng/pn-np/nb/kouchibouguac/activ
> /randonnee-hiking/tourbiere-bog.aspx

22. Miscou Island Peat Bog Boardwalk is an approximately
0.4-mi. (0.7 km) decked walkway raised over a fen that borders
a shallow pond. Interpretive signs. The island is connected by
bridges from Shippagan. The bog and boardwalk are owned by the
Canadian government and managed by the Miscou Island Tourism
Development Committee. Wheelchair accessible.

> http://www.hikingnb.ca/Trails/AcadianNorth/Miscou
> PeatBog.html

New Hampshire

23. Bradford Bog, Bradford, has an approximately 0.25-mi. or
0.4-km (one way) boardwalk consisting of two planks slightly
elevated above the surface of an Atlantic white cedar fen. The

boardwalk ends at a small observation tower overlooking a black spruce and larch wooded shrub-heath fen. Owned and managed by the Town of Bradford.

> http://nhdfl.org/events-tours-and-programs/visit-nh
> -biodiversity/bradford-bog.aspx

24. Cherry Pond and Little Cherry Pond Trail in Pondicherry Wildlife Refuge is a 7-to-9 mi. (12–15 km) hike, depending on the route. It has several short stretches of boardwalks in wooded and open acidic fens, and a viewing platform in the fen at the shoreline of each pond. Owned by the U.S. Fish and Wildlife Service and managed in collaboration with the Northwoods Stewardship Center, NH Audubon, New Hampshire Fish and Game Department, and New Hampshire Trails Bureau.

> http://www.hikenewengland.com/PondicherryNH091012
> .html

> http://www.nhdfl.org/library/pdf/Natural%20Heritage
> /Pondicherry4.pdf

25. Forsaith Forest Atlantic white cedar swamp (a wooded fen), Chester, has an approximately 0.25-mi. (0.4 km) boardwalk loop consisting of two planks blocked up above the surface. Other trees and shrubs include black tupelo, yellow birch, and coastal sweet-pepperbush. Owner and manager: Pinkerton Academy.

> http://www.nhdfl.org/events-tours-and-programs/visit-nh
> -biodiversity/ForsaithForest.aspx

26. Hurlbert Swamp, Stewartstown, is a wooded fen with abundant northern white cedar, plus other fen vegetation. An approximately 0.5-mi. (0.8 km) walk leads to a 0.2-mi. (0.3 km) boardwalk consisting of a pair of planks slightly raised above the surface. Owned and managed by the Nature Conservancy. A brochure with trail map is available at the third website.

> http://www.nature.org/ourinitiatives/regions/northamerica
> /unitedstates/newhampshire/placesweprotect/hurlbert
> -swamp.xml

> http://www.nhdfl.org/about-forests-and-lands/bureaus
> /natural-heritage-bureau/photo-index/SystemPhotos/near
> -borealsystem.aspx

http://www.nhdfl.org/library/pdf/Natural%20Heritage
/Hurlbert3.pdf

27. Loveren's Mill Preserve, Antrim, Stoddard, and Windsor, includes an Atlantic white cedar "swamp" (a wooded fen). A 2.1-mi. (3.4 km) trail (round trip) skirts the fen and has a 200-ft. (61 m) boardwalk into it consisting of a pair of planks slightly raised above the surface. A shorter option to only the boardwalk is about 0.75 mi. (1.2 km, round trip). Owned and managed by the Nature Conservancy.

http://www.nature.org/ourinitiatives/regions/northamerica
/unitedstates/newhampshire/placesweprotect/loverens
-mill-cedar-swamp.xml

http://www.nature.org/ourinitiatives/regions/northamerica
/unitedstates/newhampshire/placesweprotect/loverens
-mill-map-guide-for-web.pdf

28. Manchester Cedar Swamp Preserve, Manchester, has a 1.8-mi. (2.9 km) trail that includes a small loop in an Atlantic white cedar "swamp" (a wooded fen) where the trail becomes a boardwalk consisting of two planks slightly raised above the surface. Owned and managed by the Nature Conservancy.

http://www.nature.org/ourinitiatives/regions/northamerica
/unitedstates/newhampshire/places-preserves/manchester
-cedar-swamp-preserve.xml

29. Mud Pond Boardwalk, Fox State Forest, Hillsborough, is in an acidic fen with multiple plant communities surrounding the pond. Access to the boardwalk is by an approximately 1-mi. (1.6 km) trail. The decked and raised boardwalk is about 100 ft. (30 m) long and ends at an observation shelter near the pond shore. Owned and managed by the New Hampshire Division of Forests and Lands.

http://nhdfl.org/events-tours-and-programs/visit-nh
-biodiversity/fox-state-forest.aspx

http://www.hikenewengland.com/FoxStateForest081206.html

30. Mud Pond in Pondicherry Wildlife Refuge, Jefferson, is encircled by a fen that is accessed by an approximately 0.6-mi. (1

km) trail, the final 0.17 mi. (0.27 km) of which is a raised, decked, and railed boardwalk. The boardwalk first traverses wooded fen vegetation. As the boardwalk approaches a viewing platform near the pond shore, the fen becomes more open. Owned by the U.S. Fish and Wildlife Service and managed in collaboration with the Northwoods Stewardship Center, NH Audubon, New Hampshire Fish and Game Department, and New Hampshire Trails Bureau.

http://www.hikenewengland.com/PondicherryMudPond NH101231.html

http://www.hikenewengland.com/MapLarge-Pondicherry MudPondNH101231.html

http://www.americantrails.org/nationalrecreationtrails /trailNRT/Mud-Pond-Conte-NWR-NH01.html

31. Philbrick-Cricenti Bog, New London, is a peatland in a kettle with a variety of natural communities. It is accessed by an approximately 1-mi. (1.6 km) trail with multiple loops and boardwalks consisting of two planks slightly raised above the surface. Owned and managed by the Town of New London.

http://www.nhdfl.org/events-tours-and-programs/visit-nh -biodiversity/philbrick-cricenti-bog.aspx

http://www.nhdfl.org/library/pdf/Natural%20Heritage /Philbrick2.pdf

http://www.nl-nhcc.com/trails/philbrickcricentibogtrail %2821%29.htm

32. Ponemah Bog, Amherst, is a peatland in a kettle. It is accessed by an approximately 0.8-mi. (1.3 km) trail that includes a boardwalk consisting of two planks on the peatland surface that ends at a viewing platform near the pond shore. Owned and managed by NH Audubon.

http://hikenewengland.com/PonemahBogGen1.html

http://www.nhaudubon.org/ponemah-bog/

http://www.nhdfl.org/library/pdf/Natural%20Heritage /Ponemah3.pdf

New Jersey (Glaciated Part Only)

33. The Cedar Swamp Trail in High Point State Park, Dryden Kuser Natural Area, Montague Township, is a 2.3-mi. (3.7 km) self-guiding nature trail (trail guide available) with a 1.5-mi. (2.4 km) loop including a decked and slightly raised boardwalk at the edge of an Atlantic white cedar fen. Owned and managed by the New Jersey Park Service.

> http://www.nynjtc.org/park/high-point-state-park
> http://www.state.nj.us/dep/parksandforests/parks/highpoint
> .html#trails
> http://nynjctbotany.org/njkitt/highpont.html

New York (Glaciated Part Only)

34. Emmons Pond Bog, Davenport, occupies a kettle. It is circled by a 1.4-mi. (2.3 km) trail. At 0.2 mi. (0.3 km) along the trail a 200-ft. (60 m) decked and raised boardwalk branches off into the peatland. Owned and managed by the Nature Conservancy.

> http://www.nature.org/ourinitiatives/regions/northamerica
> /unitedstates/newyork/places-preserves/eny-emmons
> -pond-bog.xml
> http://www.nature.org/ourinitiatives/regions/northamerica
> /unitedstates/newyork/places-preserves/easternnewyork
> /wherewework/eny-emmons-map.pdf

35. Moss Lake Preserve, Houghton, has a peatland with a floating mat encircling a lake in a kettle. A 0.9-mi. (1.4 km) trail circles the peatland. If you proceed clockwise, at 0.25 mi. (0.4 m) an approximately 200-ft. (60 m) decked and raised boardwalk branches off and crosses the mat to the lakeshore. Owned and managed by the Nature Conservancy.

> http://www.nature.org/ourinitiatives/regions/northamerica
> /unitedstates/newyork/placesweprotect/centralwestern
> newyork/wherewework/central-moss-lake.xml
> http://www.nature.org/ourinitiatives/regions/northamerica
> /unitedstates/newyork/moss-lake-preserve-map.pdf

36. O. D. von Engeln Preserve, Dryden, has a 1.0-mi. (1.6 km) Bog Loop Trail with an approximately 230-ft. (70 m) decked and raised boardwalk and viewing platform in an open-wooded fen. Owned and managed by the Nature Conservancy.

http://www.nature.org/ourinitiatives/regions/northamerica /unitedstates/newyork/placesweprotect/centralwestern newyork/wherewework/central-od-von-engeln-preserve-at -malloryville.xml

http://dryden.ny.us/departments/planning-department /dryden-trails/od-von-engeln-trail

http://dryden.ny.us/Planning-Department/trails/OdVon Engeln.pdf

37. Silver Lake Bog Preserve, Hawkeye, has a 0.5-mi. (0.8 km) decked and raised boardwalk that passes through wooded and open parts of an acidic fen. A short checklist of plants and animals is available at the third website. Owned and managed by the Nature Conservancy.

http://www.nature.org/ourinitiatives/regions/northamerica /unitedstates/newyork/places-preserves/adirondacks-silver -lake-bog-preserve.xml

http://www.nature.org/ourinitiatives/regions/northamerica /unitedstates/newyork/places-preserves/adirondacks /silverlakebogpub-three-fold.pdf

http://www.nature.org/ourinitiatives/regions/northamerica /unitedstates/newyork/places-preserves/adirondacks /specieschecklistsilverlakebogblackwhite.pdf

38. Spring Pond Bog Preserve, Altamont, is a large peatland with bog and fen areas, both wooded and open. A locked gate on the access road requires advance arrangement to pass: call 518–576–2082. An approximately 0.5-mi. (0.8 km) upland trail to the bog affords excellent views of it and leads to a 375 ft. (114 m) decked boardwalk at a small cove of the main bog. Owned and managed by the Nature Conservancy.

http://www.nature.org/ourinitiatives/regions/northamerica /unitedstates/newyork/placesweprotect/adirondacks /wherewework/adirondacks-spring-pond-bog-preserve.xml

http://www.nature.org/ourinitiatives/regions/northamerica
/unitedstates/newyork/places-preserves/adirondacks
/spring-pond-bogpub-three-fold.pdf

39. Zurich Bog, Arcadia, is an acidic and circumneutral fen (the third website says it is partly a bog) with wooded and open parts in a depression of a drumlin field. It has an approximately 2-mi. (3 km) trail (round trip), including an approximately 0.3-mi. (0.5 km) boardwalk of two planks on the fen surface. Owned and managed by the Bergen Swamp Preservation Society.

Travel directions: http://www.bergenswamp.org/Maps
/Zurich_Map&Directions.pdf

Trail info and rules: http://thezurichbog.wordpress.com
/hiking-the-trails/

Geology, botany, and topographical map: http://bhort
.bh.cornell.edu/bogtrip/zurich.htm

Nova Scotia

40. Cape Breton Highlands National Park's Bog boardwalk is a 0.3-mi. (0.5 km) decked and raised walkway with interpretive signs in a "highland plateau bog" that is wheelchair accessible. It is owned and managed by Parks Canada.

http://www.pc.gc.ca/eng/pn-np/ns/cbreton/activ/randonnee
-hiking/tourbiere_bog.aspx

41. Kejimkujik National Park's Port Joli Head Trail at the Seaside is 5.4 mi. (8.7 km) long and encircles and crosses a large coastal bog. Parts of the trail consist of decked and raised boardwalks totaling about 1 mi. (1.6 km). Owned and managed by Parks Canada.

http://www.pc.gc.ca/eng/pn-np/ns/kejimkujik/activ/activ5
.aspx

42. Kejimkujik National Park's Flowing Waters Trail is a 0.6-mi. (1 km) loop with a decked and raised boardwalk in an inland bog. Owned and managed by Parks Canada.

http://www.pc.gc.ca/eng/pn-np/ns/kejimkujik/activ/activ5.aspx

Ohio (Glaciated Part Only)

43. Brown's Lake Bog State Nature Preserve, Shreve, has a kettle occupied by a peatland with wooded perimeter and open center around a small pond. An 850-ft. (260 m) boardwalk, the first 650 ft. (200 m) of which is in lowland forest, spends its final 200 ft. (60 m) in the peatland, where it ends at a viewing platform. Owned and managed by the Nature Conservancy. Second website describes a botanical tour.

> http://www.nature.org/ourinitiatives/regions/northamerica
> /unitedstates/ohio/placesweprotect/browns-lake-bog
> -preserve.xml
> http://floraofohio.blogspot.com/2014/06/browns-lake-bog
> -state-nature-preserve.html

44. Cedar Bog State Nature Preserve, Urbana, has the largest and best calcareous fen in Ohio. It contains northern white cedar, bog birch, and shrubby-cinquefoil. It is accessed by an approximately 1-mi. (1.7 km) raised boardwalk decked with four longitudinal planks and is wheelchair accessible. An Education Center runs a program, including guided walks. A field guide is available at second website. Owned by the Ohio Division of Natural Areas and Preserves (ODNAP) and the Education Center by Ohio History Connection. Managed by Cedar Bog Association and ODNAP. Entry fee.

> http://naturepreserves.ohiodnr.gov/cedarbog
> www.cedarbognp.org

45. Cooperrider (Tom S.) Kent Bog State Nature Preserve, Kent, has a 0.5-mi. (0.8 km) decked boardwalk loop at a kettle now occupied by a peatland with wooded (including American larch) and tall shrub thicket areas. Interpretive signs and guided walks. Wheelchair accessible. Owned and managed by the Ohio Division of Natural Areas and Preserves.

> http://naturepreserves.ohiodnr.gov/cooperriderkentbog
> http://naturepreserves.ohiodnr.gov/portals/dnap/pdf
> /kentbog.pdf

46. Cranberry Bog State Nature Reserve, Buckeye Lake State Park, Licking County, has a fen with an approximately 370-ft. (113

m) decked boardwalk on a floating peat island in the lake. This unusual and sensitive site is owned and managed by the Ohio Division of Natural Areas and Preserves, from which an access permit may be obtained.

> http://naturepreserves.ohiodnr.gov/cranberrybog
> http://naturepreserves.ohiodnr.gov/portals/dnap/pdf
> /cranberrybog.pdf

47. Herrick Fen State Nature Preserve, Streetsboro, has a wooded fen with American larch and an open fen with sedges and shrub-by-cinquefoil. The fen is accessed by a 1-mi. (1.6 km) trail with 0.13 mi. (0.2 km) of decked and raised boardwalk that is wheelchair accessible. Owned by the Nature Conservancy and Kent State University and managed by TNC.

> http://www.nature.org/ourinitiatives/regions/northamerica
> /unitedstates/ohio/placesweprotect/herrick-fen-nature
> -preserve.xml
> http://naturepreserves.ohiodnr.gov/herrickfen
> http://www.nativetreesociety.org/fieldtrips/ohio/herrick
> _kent.htm

48. Jackson Bog State Nature Preserve is in Massillon. Jackson Bog is really an alkaline fen. The 1.25-mi. (2.1 km) trail, about half of which is decked boardwalk, has interpretive signs. Owned and managed by the Jackson Township Local Board of Education and the Ohio Division of Natural Areas and Preserves.

> http://naturepreserves.ohiodnr.gov/jacksonbog

Ontario (North to 49° N Latitude)

49. Alfred Bog Walk, Alfred, is a 0.3 mi. (0.5 km) trail with 0.17 mi. (0.28 km) of decked boardwalk at the minerotrophic periphery of a large raised bog. Owned by the Nature Conservancy of Canada (NCC) and managed by Ontario Parks in collaboration with NCC as a Provincial Nature Reserve.

> http://www.ofnc.ca/conservation/alfredbog/index.php
> http://www.nation.on.ca/recreation/hikingwalking/alfred
> -bog-walk

https://willowhousechronicles.wordpress.com/2009/06/13
/alfred-bog/

50. Cold Creek Conservation Area, Nobleton, has a 0.5-mi. (0.8 km) raised, decked, and railed boardwalk in a mixed conifer fen with black spruce. Environmental education programs. Owned by the Toronto and Region Conservation Authority (TRCA) and managed by the Cold Creek Stewardship, King Township, and the TRCA.

http://coldcreek.ca/

http://coldcreek.ca/wp-content/CCCA-TH.pdf

51. Mer Bleue Conservation Area, Ottawa, contains the exceptional Mer Bleue Bog, a Ramsar Site. The large raised bog has a 0.75-mi. (1.2 km) self-guided nature trail (brochure at second website), most of which is a decked and raised boardwalk, through open fen and open and wooded shrub heath bog communities. Interpretive signs. Wheelchair accessible. Located in Canada's Capital Greenbelt, it is owned and managed by the National Capital Commission.

http://www.ncc-ccn.gc.ca/places-to-visit/greenbelt/mer-bleue

http://www.ncc-ccn.gc.ca/sites/default/files/pubs/NCC-Mer
-Bleue-Bog-Special-Place.pdf

https://willowhousechronicles.wordpress.com/2009/08/17
/visiting-mer-bleue/

52. Sifton Bog, London, is a peatland with a central pond in a kettle. A 0.22-mi. (0.37 km) decked boardwalk ends at a viewing platform at the edge of the pond. Owned and managed by the Upper Thames River Conservation Authority and the City of London.

http://thamesriver.on.ca/parks-recreation-natural-areas
/londons-esas/sifton-bog/

http://thamesriver.on.ca/wp-content/uploads/ESAs/2011
_siftonbog_brochure.pdf

53. Spruce Bog Boardwalk Trail, in Algonquin Provincial Park near Huntsville, is a 0.93 mi. (1.5 km) loop trail with stretches of planked and raised boardwalk in two separate peatlands, an acidic stream-valley fen, and a much smaller peatland in a kettle. Wheelchair accessible. Printed guides keyed to numbered posts along the trail may be purchased by link from the following website.

Owned and managed by the province of Ontario.

http://www.algonquinpark.on.ca/visit/recreational_activites
/spruce-bog-boardwalk-trail.php

54. Wainfleet Bog Conservation Area, Wainfleet, has an approximately 1.25-mi. (2 km) trail including three stretches of boardwalk totaling about 0.25 mi. (0.4 km) and consisting of three planks raised above the surface in a large raised bog. The bog has been damaged by drainage and peat extraction. Restoration activities are under way. Owned and managed by the Niagara Peninsula Conservation Authority and the Nature Conservancy of Canada.

http://www.npca.ca/conservation-areas/wainfleet-bog/
http://files.ontario.ca/environment-and-energy/parks-and
-protected-areas/mnroo_bcr0232.pdf

Pennsylvania (Glaciated Part Only)

55. Nuangola Bog, Nuangola, is a small scrub-shrub acidic fen bordered by a wooded fen. It has two stretches of decked boardwalk on the peatland surface, the north stretch about 530 ft. (160 m) long, and the west stretch about 500 ft. (150 m) long. The peatland occupies a kettle together with adjacent Nuangola Lake. Owned by Wilkes University and managed by the Nature Conservancy, with boardwalk maintenance by the Nuangola Bog Lakefront Property Owners Association.

http://klemow.wilkes.edu/Nuangola-Bog.html

56. Tannersville Cranberry Bog is an acidic fen in a kettle at Tannersville near the limit of the Wisconsinan glaciation. An approximately 0.9-mi. (1.5 km) trail includes a 0.25-mi. (0.4 km) floating, decked boardwalk. Owned and managed by the Nature Conservancy, with the collaboration of the Kettle Creek Environmental Education Center (KCEEC). Visits only by scheduled or prearranged KCEEC guided walks (see second website).

http://www.nature.org/ourinitiatives/regions/northamerica
/unitedstates/pennsylvania/placesweprotect/tannersville
-cranberry-bog-preserve.xml#thingsToDo

http://www.mcconservation.org/downloads/PDFs/2015
%20bog%20program%20schedule%20QR%20KC.pdf
57. Thomas Darling Preserve at Two-Mile Run, Blakeslee, has an approximately 2.2-mi. (3.7 km) loop trail with sections of decked and planked boardwalk in a mosaic of wetlands including fens. Owned and managed by the Nature Conservancy and the Wildlands Conservancy of Tobyhanna Township.

http://www.nature.org/ourinitiatives/regions/northamerica
/unitedstates/pennsylvania/placesweprotect/thomas
-darling-preserve-at-two-mile-run.xml

Prince Edward Island

58. Ellerslie Bog Walk, adjacent to the Confederation Trail between Ellerslie and McNeills Mills, is a 1.4-mi. (2.4 km) trail with planked boardwalk ending at an interpretive deck in a small raised bog. Owned and managed by the province.

http://www.gov.pe.ca/infopei/index.php3?number
=65370&lang=E
59. North Cape or Black Marsh Nature Trail, at North Cape Bog or Black Marsh, is 2.1 mi. (3.5 km) long and includes an approximately 0.6-mi. (1 km) planked boardwalk ending at a viewing platform at the approximately 400-acre (160 ha) raised bog. Interpretive signs. Wheelchair accessible. Owned by the province and managed by the Tignish Initiative.

http://islandtrails.ca/en/trail.php?t=6
http://www.northcape.ca/black-marsh-nature-trail/

Québec (North to 49° N Latitude)

60. Bordelais Bog Peatland Boardwalk, in Parc naturel de la Tourbière-du-Bordelais, near a residential area of Ville de Saint-Lazare, is a 0.4-mi. (6.7 km) decked and raised interpretive boardwalk in a small fen. Owned and managed by Ville de Saint-Lazare.
61. Johnville Bog and Forest Park, Cookshire-Eaton, has a 3.5-

mi. (5.8 km) interpretive trail system including an approximately 1.5-mi. (2.4 km) bog route with about 0.4 mi. (0.6 km) of longitudinally decked boardwalk on the peatland surface. Owned by the University of Sherbrooke and Bishop's University and managed by Nature Eastern Townships.

http://www.bonjourquebec.com/fr/pdf/regional-municipal-park-johnville-bog-and-forest-park-375443535.pdf

http://provincequebec.com/forests-parc-mountains-quebec/johnville-park/

62. Le Marais de la Rivière aux Cerises, Magog, has the approximately 0.7-mi (1.1 km) Petit Houx Trail that is mostly a decked and raised boardwalk. It traverses a variety of wetlands, including fens. There is an associated interpretive center, Le centre d'interprétation du Marais. The first two of the following websites are in French. The third is an English machine translation of the second website.

http://maraisauxcerises.com/files/SentiersMarais_Depliant2011.pdf

http://maraisauxcerises.com/

http://www.microsofttranslator.com/bv.aspx?ref=SERP&br=ro&mkt=en-US&dl=en&lp=FR_EN&a=http%3a%2f%2fmaraisauxcerises.com%2f

63. Marlington Bog, Ogden, is reached by an approximately 0.7-mi. (1.1 km) trail from the Davis Road. The second half of the trail has sections of planked boardwalk in a cedar swamp or fen. The largely open and acidic main fen has two lengths of boardwalk totaling about 0.16 mi. or 0.26 km, and consisting of two planks on crosspieces, one length ending at a viewing platform near a central pond. Owned by the Nature Conservancy and managed by Conservation Elliandress.

http://townshipsheritage.com/article/marlington-bog

64. La Tourbière, in Frontenac National Park, has an approximately 1.3-mi. or 2.2-km (one way) decked interpretive boardwalk largely through an acidic wooded fen. The final 0.12 mi. (0.2 km) is in an open patterned (ribbed) fen and ends at an observation tower. Owned by Parcs Québec and managed by Société des étab-

lissements de plein air du Québec.

http://www.easterntownships.org/hikingTrails/21/la-tourbiere
http://www.sepaq.com/pq/fro/information.dot?language_id=
65. La Tourbière hiking trail in La Mauricie National Park, near Shawinigan, is about 0.2 mi. (0.3 km) long and has a 0.15-mi. (0.24 km) decked boardwalk that loops through a lakeside acidic fen. Owned and managed by Parks Canada.

http://www.pc.gc.ca/eng/pn-np/qc/mauricie/activ/activ6
.aspx#EasyHiking
66. Tourbières-de-Lanoraie Ecological Reserve, 6.2 mi. (10 km) southeast of Joliette, has an approximately 0.4-mi. (0.65 km) decked boardwalk on the surface of an acidic fen. Owned and managed by the Québec government. The first website is in French; the second is an English machine translation of it.

http://www.mddelcc.gouv.qc.ca/biodiversite/reserves
/tourbieres_lanoraie/res_48.htm
http://www.microsofttranslator.com/bv.aspx?ref=SERP&br
=ro&mkt=en-US&dl=en&lp=FR_EN&a=http%3a%2f
%2fwww.mddelcc.gouv.qc.ca%2fbiodiversite%2freserves
%2ftourbieres_lanoraie%2fres_48.htm
67. Pointe-Taillon National Park, near Saint-Henri-de-Taillon, has an approximately 0.12-mi. (0.2 km) raised boardwalk decked by four longitudinal planks in a small part of the extensive fen that characterizes most of the park. It is located near km 1 of the main bike trail leading from the visitor center. Naturalist program. The website contains a link to a visitor guide with map. Owned and managed by Parcs Québec.

http://www.sepaq.com/pq/pta/index.dot?language_id=1

Vermont

68. Chickering Bog Natural Area, East Montpelier and Calais, has an approximately 100-ft. (30 m) decked boardwalk in a circumneutral sedge fen in a bedrock basin. An approximately 1-mi. (1.7 km) trail leads to the boardwalk. Owned and managed by the Nature Conservancy.

http://www.nature.org/ourinitiatives/regions/northamerica
/unitedstates/vermont/placesweprotect/chickering-bog
-natural-area.xml

69. Colchester Bog Natural Area, Colchester, has an approximately 350-ft. (107 m) raised and decked boardwalk leading into a largely wooded fen. Viewing platform at end. Owned and managed by the University of Vermont.

http://www.uvm.edu/envprog/natural-areas/colchester
-bog

70. Eshqua Bog, in Eshqua Bog Natural Area, Hartland, is a small circumneutral fen rich in wildflowers and dragonflies. A 460-ft. (140 m) decked boardwalk in the fen is part of a 1-mi. (1.6 km) trail system. Eshqua Bog Natural Area is co-owned and managed by the Nature Conservancy and the New England Wildflower Society. Wheelchair accessible.

http://www.nature.org/ourinitiatives/regions/northamerica
/unitedstates/vermont/placesweprotect/eshqua-bog-natural
-area.xml

71. Lake Carmi Bog, in Lake Carmi State Park, Franklin, has a raised and decked 230-ft. (70 m) boardwalk loop in the largely wooded bog. Owned and managed by the Vermont Department of Forests, Parks, and Recreation.

http://www.vtstateparks.com/htm/carmi.htm
http://vtstateparks.blogspot.com/2012_11_01_archive.html

72. Mollie Beattie Bog Boardwalk, near Island Pond in the Nulhegan Basin Division of the Silvio O. Conte National Fish & Wildlife Refuge, is an approximately 200-ft. (60 m) raised and decked walkway that ends at an observation platform in the wooded peatland. It has interpretive signs and is wheelchair accessible. Owned and managed by the U.S. Fish and Wildlife Service.

http://www.islandpond.com/nature/index.htm
http://www.fws.gov/refuges/trails/refuge.cfm?orgcode=53590

73. Moose Bog Trail, in Wenlock Wildlife Management Area, Ferdinand, is about 0.8 mi. (1.3 km) long, including an approximately 0.1-mi. (0.15 km) spur trail into the acidic fen surrounding Moose Bog Pond. The spur trail is a boardwalk consisting of two

planks, side by side. Owned and managed by the Vermont Fish and Wildlife Department.

> http://www.vtfishandwildlife.com/UserFiles/Servers/Server
> _73079/File/Where%20to%20Hunt/St.%20Johnsbury
> %20District/Wenlock%20WMA.pdf
> https://www.youtube.com/watch?v=q-4XqygZ57Y

Wisconsin (Glaciated Part Only)

74. Beulah Bog, Walworth County, is a wooded bog with American larch in a kettle. A 0.35-mi. (0.6 km) trail leads to the bog edge, where decked bridging crosses the lagg. From there, an approximately 330-ft. (100 m) two-planked boardwalk leads to a narrow open mat at pond's edge. The planks may be flooded. Owned and managed by the Wisconsin Department of Natural Resources.

> http://www.wisconsinwetlands.org/Gems/SE1_Beulah_Bog
> .pdf
> http://dnr.wi.gov/topic/Lands/naturalareas/index.asp?SNA
> =122

75. Brule Bog Boardwalk Trail, in the Brule Glacial Spillway State Natural Area just northeast of Solon Springs, traverses uplands, swamps, and wooded fens. It is 2.3 mi. (3.8 km) long and has about 0.7 mi. (1.1 km) of raised and decked boardwalks that are wheelchair accessible in parts (see first website). Owned and managed by the Wisconsin Department of Natural Resources.

> http://northcountrytrail.org/bsc/wp-content/uploads/2012
> /10/bruleguide.pdf
> http://wisconsinwetlands.org/Gems/NW5_Brule_Glacial
> _Spillway.pdf

76. Cedarburg Bog Natural Area, Saukville, has a 0.25-mi. (0.4 km) access trail to a 0.65-mi. (1.1 km) boardwalk of variable structure through largely wooded circumneutral fen communities, and ends at what may be the southernmost patterned or ribbed fen in North America. A natural history guide published as *Field Station Bulletin*, vol. 18, no. 2, may be downloaded from the fourth website. Owned by the University of Wisconsin (UW) and the Wis-

consin Department of Natural Resources (WDNR) and managed by the UW-Milwaukee Field Station (fourth website), WDNR, and the Friends of the Cedarburg Bog. Advanced permission for entry to the boardwalk may be obtained from the field station, 262–675–6844, or fieldstn@uwm.edu.

http://www.wisconsinwetlands.org/Gems/SE2_Cedarburg _Bog.pdf

http://dnr.wi.gov/topic/Lands/naturalareas/index.asp?SNA=2

http://bogfriends.org/

https://www4.uwm.edu/fieldstation/

77. Forest Lodge Nature Trail, Cable, is an approximately 1.2-mi. (2 km) trail with interpretive stations and a booklet. The trail becomes an approximately 300-ft. (90 m) decked and raised boardwalk where it crosses an open-wooded fen with American larch. Owned and managed by the Chequamegon-Nicolet National Forest and the Cable Natural History Museum.

http://wisconsintrailguide.com/hiking/forest-lodge-nature -trail.html

78. Spruce Lake Bog State Natural Area, at the Northern Unit of Kettle Moraine State Forest near Campbellsport, has an approximately 575-ft. (175 m) trail, the final two-thirds of which is a three-planked boardwalk loop through a circumneutral wooded fen with black spruce and American larch, and through an open floating mat bordering a kettle lake. Owned and managed by the Wisconsin Department of Natural Resources.

http://www.wisconsinwetlands.org/Gems/SE11_Spruce _Lake_Bog.pdf

http://dnr.wi.gov/topic/Lands/naturalareas/index .asp?SNA=59

Index of Subjects
and Common Plant Names

Page numbers in *italics* refer to illustrations. Species in bold type are featured and illustrated.

Index of Latin Plant Names and Families

Page numbers in *italics* refer to illustrations. Species in bold print are featured and illustrated.